TOOLS AND TACTICS OF DESIGN

TOOLS AND TACTICS OF DESIGN

PETER G. DOMINICK
Assessment Alternatives, Inc.

JOHN T. DEMEL
The Ohio State University

WILLIAM M. LAWBAUGH
Mount Saint Mary's College

RICHARD J. FREULER
The Ohio State University

GARY L. KINZEL
The Ohio State University

ELI FROMM
Drexel University

JOHN WILEY & SONS, INC.
New York / Chichester / Weinheim / Brisbane / Singapore / Toronto

Acquisitions Editor *Joseph Hayton*
Editorial Assistant *Steven Peterson*
Marketing Manager *Katherine Hepburn*
Production Services Manager *Jeanine Furino*
Production Editor *Sandra Russell*
Senior Designer *Karin G. Kincheloe*
Illustration Editor *Eugene Aiello*
Photo Editor *Nicole Horlacher*
Production Management Services *TechBooks*
Cover Image © *Photo Disc, Inc.*

This book was set in 10/12 *Times Roman* by *TechBooks* and printed and bound by *Courier Companies* (*Westford*). The cover was printed by *Phoenix Color Corporation.*

This book is printed on acid-free paper. ⊗

Library of Congress Cataloging in Publication Data:
Tools and tactics of design / Peter G. Dominick . . . [et al.].
Includes bibliographical references and index.
ISBN 0-471-38648-0 (pbk. : alk. paper)
1. Engineering design. I. Dominick, Peter G.

TA174 .T66 2001
620′.0042—dc21 00-043298

Printed in the United States of America

10 9 8 7 6 5 4 3 2 1

PREFACE

This book describes the engineering design process. It introduces the key steps and the activities associated with those steps. In this text, however, our objective is to do more than just tell you about design. We want to provide you with a resource that will support your own design experiences. The following pages contain tools, tactics, and other resources that are best learned by applying them to your own design projects. On the one hand, this text serves as a general resource for learning about the design process. On the other hand, it is a specific reference that you can apply to your own design work. Although much of what we discuss applies to design projects undertaken by engineering professionals, it has been written specifically with student design projects in mind. Therefore, we hope this book is something you will use rather than just something you will read.

As you put the text into practice, keep in mind three key principles that have guided our work. First of all, the fundamental characteristics of engineering design transcend any specific engineering discipline. These characteristics include issues like clear problem definition, a systematic search for solutions, and the importance of rigorous testing.

Secondly, engineering design is not just a thing. It is a personal process that involves you and the people with whom you work on your design project(s). In other words, the best way to introduce and describe engineering design is through the thoughts and actions of the people who are doing it. We have chosen to describe these thoughts and actions in terms of four broad skill areas that apply to design and to problem solving in general: decision making, project management, communication, and collaboration. Understanding your own strengths and improvement areas in relation to these skills will help you to become better at engineering design. Using this text will provide you with insights to your abilities. More importantly, this text can guide you in developing your own skill level.

The third related principle guiding our work is that design is best learned by experiencing it. Therefore, we think that a textbook about engineering design must do more than just describe. It has to help you get the job done. We want you to see the implications of your work and learn from your mistakes as well as your successes. The only way to do so is through hands-on experience.

We are by no means solely responsible for the writing of this book. Its origins are derived from the work being done by numerous faculty who are part of the Gateway Engineering Education Coalition. Founded in 1992, the Coalition is headquartered at Drexel University and currently consists of engineering faculty from Columbia University, the Cooper Union for the Advancement of Science and Art, Drexel University, the New Jersey Institute of Technology, the Ohio State University, the Polytechnic Institute of New York, and the University of South Carolina. As our Coalition's mission statement puts it, "This Coalition is opening new gateways for learning by altering engineering education from a focus on course content to a focus on the broader experience in which individual curriculum parts are connected and integrated." Since its founding, hands-on design experiences have been an integral aspect of the Coalition's efforts to connect and integrate curriculum content. At Drexel University, for instance, freshman students work in teams on year-long design projects of their own choosing. They are responsible for everything, including the identification and selection of a problem. Similarly, at the Ohio State University, students work on an open-ended design project that transcends several terms. At the Cooper Union, instructors pose open-ended engineering problems currently being faced by the City of New York.

The instructors at these schools (and the Coalition in general) believe that these kinds of experiences are crucial to engineering education. This is not to say that traditional emphases on science and math are not also important to an engineer. Of course they are. But experientially oriented design projects help new students place those basic fundamentals into context. This context enables students to appreciate what it means to be an engineer and how the profession is unique.

In addition to Gateway, the Accreditation Board for Engineering and Technology (ABET) also influenced our approach to this book. ABET's new Outcome Criteria for assessing the effectiveness of engineering colleges requires schools to demonstrate that students are not only technically competent but also are skilled in other important professional areas like teamwork and communication. In many cases, teaching these vital skills is beyond the scope of engineering instructors who understandably have to focus on their technical specialties. We wanted to write a text that would support their efforts to teach these softer skills.

In closing, we are compelled to make one more observation that helped to confirm, for us, the value of the experiential approach to teaching design. As we came together to work on this project, it did not take long to realize that writing this book was itself a typical open-ended design project. We encountered many of the issues we were trying to write about. Along the way we went through several iterations of our ideas and concepts, and at times we struggled to define our goals and objectives more clearly. We also dealt with project management issues like determining who should do what, meeting deadlines (our editor will attest to the challenges we faced in this regard), and how to review our work.

Coming together as a team was another challenge we faced, partly because we were dispersed geographically. In addition, we came from diverse professional backgrounds. Among us were academics and practitioners, engineers, an industrial psychologist, and a communications expert. This diversity was ultimately a strength and enabled us to learn from one another. On the other hand, it also posed some interesting communication challenges. For instance, a psychologist's definition of a process can be quite different from an engineer's definition of the same term. (You will learn how both perspectives relate to design.) Like most teams, our development took time, and we all had to surrender some of our individual aims for the overall benefit of the project. All these issues and more are what design teams must face. We are glad to have had the chance to face them and hope you benefit from the result.

ACKNOWLEDGMENTS

In writing this book we were very fortunate to have benefited from the support, input, and criticisms of many people. We want to acknowledge their invaluable efforts and express our gratitude. Jack McGourty (Columbia University and Gateway) was instrumental in putting our team together and also helped to shape the skill-based approach we use to explain design. Veona Martin (Assessment Alternatives Inc.) and Sandra Lionetti (LA Dreyfus Company) made essential contributions to our discussions of team skills. In particular, they helped prepare material you will read on collaboration and communication skills. We are also thankful for the contributions and ideas provided by several people at the National Aeronautics and Space Administration (NASA). They are Frank Hoban, Pat Patterson, Denny Van Liere, Tommy Kirk, Richard Evans, Ed Hoffman, and Tony Maturo. Thanks also go to our editor, Joe Hayton of John Wiley & Sons. His consistent support and occasional urging kept us working towards our deadlines.

As mentioned in the preface, this book is the outgrowth of work undertaken by faculty who are part of the Gateway Coalition. We are indebted to all of them and to their students as well. In particular, we want to thank Audeen Fentiman (the Ohio State University), Jean LeMee (the Cooper Union for the Advancement of Science and Art), Dick Weggel (Drexel University), and Tony Renshaw (Columbia University).

GATEWAY ENGINEERING EDUCATION COALITION

The Gateway Engineering Education Coalition is a collaborative program of seven institutions, supported by the Engineering Directorate of the National Science Foundation. Headquartered at Drexel University and representing a diversity of institutional cultures, the Coalition is opening new gateways for learning by shifting undergraduate engineering education to include a focus on the development of human resources and the broader experience in which individual curriculum and professional parts are connected and integrated as well as its traditional focus on course content. The institutional members of the Coalition include the Schools of Engineering of Columbia University, the Cooper Union, Drexel University, New Jersey Institute of Technology, Ohio State University, Polytechnic University of New York, and the University of South Carolina. To accomplish their common held objectives the institutions have modified the structure as well as the content of their undergraduate engineering programs to emphasize the entire educational process and the interconnectedness of its many parts, not just individual courses.

The intellectual threads weave together engineering-up-front, the supporting knowledge base of unified and connected science and math concurrently with engineering, the integrative and professional aspects of the engineering process, the importance of communication skills, teaming, and leadership skills with a multidisciplinary emphasis, and the use of technologies to enhance the educational process in the development of the emerging engineering professional of the 21st century.

The Coalition has had focus themes to address a broad range of issues pertaining to the educational process. These include curriculum innovation in structure and content; professional development for students and faculty; instructional technologies; programs to enhance continued participation of populations traditionally under-represented in engineering; linking and sharing across traditional boundaries among departments, colleges, and institutions; and an assessment and feedback program to ascertain the effectiveness and continual improvement of these programs. Design, as a motivator, facilitator, as well as an area of study, has had an important impact on many of the focus themes of the Gateway program. It has its impact, of course, in the curricular aspect directly. In addition, however, design has become a vehicle to address the issues of professional development, communication skills development, leadership skills development, and teaming, not only among long time colleagues but also colleagues across institutions. Design also supports a better understanding of different institutional and individual cultures. Thus, design has become an important player in developing the technical, scientific, professional, and personal attributes of our emerging engineering professionals. This text, *Tools and Tactics of Design*, highlights those multiple facets of the engineering educational process.

This work has been sponsored by the Gateway Engineering Education Coalition and supported, in part, by the Education and Centers Division of the Engineering Directorate of the National Science Foundation (Awards EEC-9109704 and EEC-9727413).

CONTENTS

TOOLS AND TACTICS OF DESIGN

CHAPTER 1

INTRODUCTION

1.1 INTRODUCTION TO ENGINEERING DESIGN—THE ART OF DESIGN

Imagine life without the following few selected contemporary items: Cellular phones and the network of cells for transmitting sound, electric toothbrushes and the method for getting toothpaste into a tube, portable CD players and the procedures for making CDs, disposable cameras and the chemical process for developing film, modern mountain bikes and bicycle trails, plastic soft drink bottles and their carbonated contents, space stations and the life support systems enabling them to be built. Each of these products, processes, and systems represents major design advancements of the recent past. All were developed by modern design theory where the objective is to bring products to market or create a process that is high in quality, reliable, competitive in terms of cost, and available in a timely fashion.

Design is exciting because you are creating things that did not exist before. In fact, virtually everything not created by nature is designed. Clothes are designed. Artwork is designed. New buildings are designed. Road systems are designed. What does this mean and where does engineering fit in?

In essence engineers are designers. They figure out ways to provide the things that are needed or wanted. They do this by inventing or designing something that was never there before. Many say that the process of designing is what distinguishes the profession of engineering from pure science. In this book and in your engineering courses, you will learn about the engineering design process and how to apply it.

A product or process can be initially designed or initially invented and meet the basic need or desire of the customers. Through additional design (often called development), the product or process can be improved—made cheaper, easier to use, more comfortable, more reliable, and of higher quality. For instance, the first iterations of a new invention or model may include very basic components with a focus being on just getting the design to work. When you add analysis to the design process, optimum levels of power, comfort, shape, size, and functionality may be added. Your engineering design contribution brings a quantitative perspective to the creation of new things to enhance the creative or qualitative perspective.

Individual inventors and designers are remembered for their creations but those individually created items and the individuals tend to be from a past era. We think of Henry Ford, Alexander Graham Bell, and Thomas Edison when we think of the car, the telephone, and the electric light bulb. In the old days, the term inventor implied one person working alone (Figure 1-1). In reality, however, even these three "rugged individualists" recognized the importance of collaborative efforts. In particular, Thomas

Henry Ford
Transmission Mechanism
Patent No. 1,005,186

Alexander Graham Bell
Telegraphy
Patent No. 174,465

Thomas Alva Edison
Electric Lamp
Patent No. 223,898

FIGURE 1.1 Early individual inventors and their products of the past. [(left): Underwood & Underwood/CORBIS. (center): ©CORBIS. (right): Courtesy Edison Natural History Society.]

Edison is considered by many historians to be responsible for institutionalizing the process of research, development, and invention by teams.

If engineering design ever was a solitary endeavor it certainly is not today. Most new products, systems, and processes are produced by teams. This means that each engineer must be able to work with other engineers, scientists, business managers, communications experts, and industrial designers. As the examples in Figures 1.2 and 1.3 imply, being a engineer in the 21st century also means being part of a team.

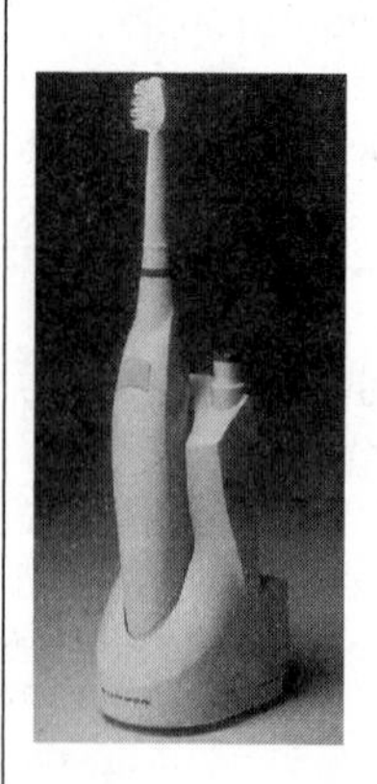

Patent Number: 5,613,259

This electronic toothbrush was designed by a cross-functional and colocated product development team, consisting of engineers and scientists with backgrounds in mechanical and electrical design, polymers, magnetics, dynamics, and manufacturing. The design team scrapped the use of a traditional motor that relies on gears or pistons to drive the bristles in favor of a more efficient motor called a resonant mechanical oscillator. This new motor delivered more energy to the end of the toothbrush, resulting in better plaque removal. The product debuted just 13 months after the project began.

Inventors: *Craft; Adam B.* (Fort Collins, CO); *Schleiffer; Keith E.* (Gahanna, OH); *Dvorsky; James E.* (Hilliard, OH); *Graves; Thomas W.* (Fort Collins, CO); *Gray, III; Ronald.* (Columbus, OH); *Senapati; Nagabhusan* (Worthington, OH); *Zelinski; Matthew S.* (Worthington, OH)

FIGURE 1.2 An electronic toothbrush developed by a design team. (Courtesy Battele Memorial Institute.)

This enteral feeding pump was also designed by a cross-functional and colocated product development team in Ohio, consisting of engineers and scientists with backgrounds in mechanical and electrical design, industrial design, polymers, software, and manufacturing. An enteral feeding pump is a device for providing nutrition to people who cannot or will not consume food orally. The design team developed a pump that met a stringent unit manufacturing cost with improved technical features that was easier to use and to maintain.

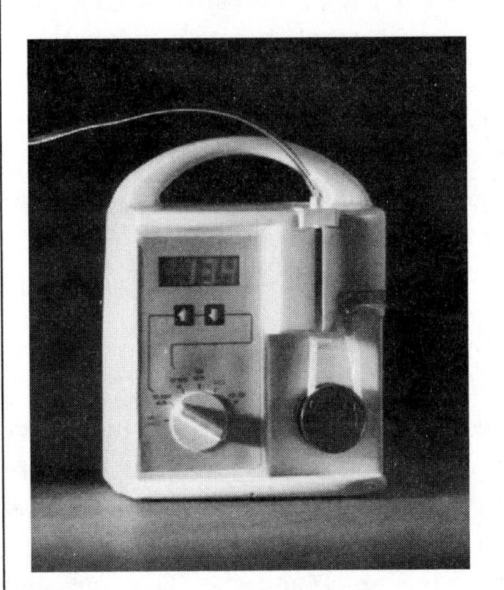

Patent Number: 5,807,333

Ohio Project Team (***from left to right***): *Clark Fortney, W. Fred Lyon, Cathy Alexander, Grant Wilson, Patina Ripkey, Bill Patton, Doug Vrona, Matt Fleming, Joe Juratovac,* and *Dennis Hoffman*

FIGURE 1.3 An enteral feeding pump developed by a design team. (Courtesy Battele Memorial Institute.)

This book addresses the engineering design process, how to manage the design process in an academic environment, and how to work effectively in a design project team. Good ideas, precise analysis, and good quality control must be communicated effectively among the team members and the sponsors if the design process is to be executed optimally.

The next part of this chapter introduces the organization of this text and describes our approach to explaining design. The third part highlights additional characteristics of the design process and defines some terms with which you should be familiar.

1.2 ORGANIZATION OF THIS DESIGN TEXT

If you are using this text for the first time, you are most likely embarking on some kind of introductory design experience. Along the way, you will be learning first-hand what the art and science of engineering design is all about. This text will assist you as you grapple with the engineering design process and the skills it requires. Those who have some prior design experience may want to use portions of the text more selectively, but there is something of value to improve the engineering design process for everyone here.

As you read this text, you will find different design projects that are used as examples. These include a freshman robot design project and a robotic arm attached to wheelchairs as an assistive feeding device. The robot design project is done by first-year students at Ohio State University and by second- and third-year students at MIT. The robotic arm project has been done by seniors in a collaboration that includes Cooper Union, Drexel University, New Jersey Institute of Technology, and Ohio State. Work on such projects is performed by a team of faculty, research engineers, and graduate and undergraduate students. In addition, there will be one continuing project, the redesign of a dorm room that allows you and your teammates to try out the elements of the design process.

1.2.1 Intersection of Phases and Steps with Skills and Activities

As you read about them, you will see that each of the design projects mentioned above has unique characteristics. On the other hand you should also recognize that regardless of what is being designed, there are common features in the design process. For instance, all design efforts are about systematic problem solving. They are all cyclical and iterative, and they all have a finite beginning and end. There are also some fundamental skills that all designers must develop regardless of what they are designing.

This is not to say that all design is identical across all engineering disciplines. For instance, civil engineering design for construction is different from that of a mechanical engineer engaged in product design. For the civil engineer, there is usually a request for project bids with a set of project specifications. The civil engineer or engineering firm estimates what the engineering design work will cost and submits the bid to a contracting agency. If the bid is accepted, the agency writes a contract with the engineer or engineering firm. Once the contract has been duly signed and witnessed, the actual engineering design work is done. The individual or firm provides a set of drawings and written specifications for the project to the contracting agency. If no changes are required, the agency requests bids for the actual construction work from a construction company. The engineer or engineering firm usually provides engineering oversight for the project and updates the design drawings and details. The contracting agency gets a series of "as-built" drawings that document what was actually constructed.

In contrast, the chemical engineer is usually engaged in designing a process to produce chemicals, petroleum products, food items, and other commercial materials. This requires testing all the process components and may require building a pilot plant to see if the process really works. If it does, then a full-scale plant is constructed. Again, the chemical engineers who did the design may supervise construction and do the testing of the completed plant.

In dealing with product design, design for construction, and process design, there are clearly differences in terms of technical and scientific skills required to do one instead of the other. These distinctions are important, and you will learn more about them during your engineering education.

Our focus here is on those aspects of engineering design that are common across all engineering disciplines. Specifically, our approach to supporting your design efforts intersects two related ways of thinking about the design process. First, all design processes consist of several iterative phases. We have defined these phases as defining the problem, formulating solutions, developing models and prototypes, and presenting the design.

In addition to thinking about the phases in the design process, it is equally important to understand that *you* are the agent who makes design happen. In other words, another way of thinking about the process of design would be in terms of *your* thoughts and *your* actions. Therefore, design can be characterized by your behavior in relation to four broad skill areas: decision making, project management, communication, and collaboration.

1.2.2 Phases and Steps of the Design Process

We have chosen to break the design process into four phases because they best reflect the design steps that most students encounter in introductory design courses. As you review our definitions for these phases, keep in mind that all design processes have a beginning and an end but in between the process is iterative. To some extent, you will be constantly defining and redefining problems, formulating and reformulating solutions, and building and revising your design concepts as you work your way through the experience of the engineering design process. The boxes in Figure 1.4 provide a more detailed picture of each phase of the design process. The lines with arrows connecting these boxes provide an indication of how the design process is iterative.

Phase One: Defining the Problem Defining the problem means clearly exploring and articulating the nature of the problem to be solved. This is the first and often the most difficult part of the design process. You will learn that an engineer's definition of a problem must reconcile two opposite tendencies. On the one hand, the problem must be fully clarified and articulated in terms that are as definite as possible. On the other hand, the definition must remain open-ended enough so that it does not preclude consideration of all feasible solutions. The kinds of questions you must consider when defining a problem include the following:

- What requirements must our design meet?
- What solutions to the problem currently exist?
- What are the constraints to these solutions (technical, economic, social, political, environmental, etc.)?

Moreover, you will learn that you must define your problem in clear and succinct terms that can be understood not just by other engineers but also by those who

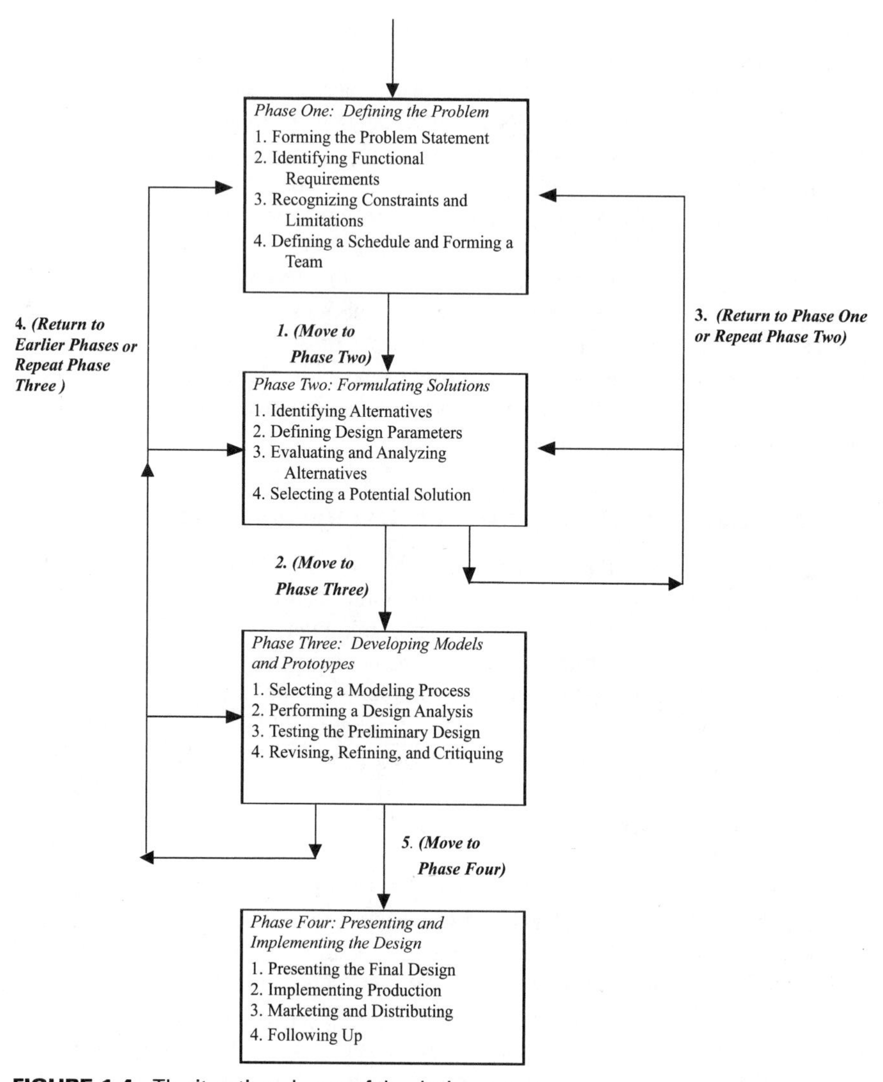

FIGURE 1.4 The iterative phases of the design process.

will benefit from your work. Remember that you will probably be working as a member of a concurrent-design team. By concurrent, we mean that most design teams in industry include professionals from other disciplines such as marketing, manufacturing, quality control, and frequently customers. These other team members may or may not have an engineering background but they are all stakeholders in the design process. Bringing people together concurrently improves coordination and understanding across functions. For instance, it helps to ensure that designers do not create something the manufacturers cannot realistically produce. You will learn more about the concurrent design process in Section 1.3.6.

Phase Two: Formulating Solutions Formulating solutions centers on systematically exploring and evaluating alternatives in relation to the requirements of the design problem. You will learn that there is typically more than one solution to your design problem. Your challenges in this phase include first identifying as many of these alternatives as possible and then assessing their pros and cons in order to determine the optimal solution. When formulating solutions, engineers typically have to make tradeoffs between various design parameters. In other words, an ideal solution for one part of your design problem may confound solutions to other parts of your problem. This text will provide you with some guidelines for evaluating your design efficiently and systematically. It will also help you answer questions such as: What kinds of data should we consider? What skills will we need? What kinds of analyses must we do to demonstrate that our solution will solve the problem? As a result, you will be in a better position to justify the rationale behind the solution that you have chosen.

Phase Three: Developing Models and Prototypes This phase of the design process involves translating design solutions into tangible outputs that demonstrate and help to refine the feasibility of your solution. In this section of the text, you will learn about how to select the right kind of modeling process for your design, how to perform a design analysis, how to test your design, and finally how to revise and critique it. This is truly the hands-on part of the design process. Modeling brings your design to life and helps you gain a better understanding of its eventual effectiveness. Like all aspects of design, you will find this phase to be quite iterative. As you develop your model, you may find yourself going back to revise your alternatives or perhaps even to redefine your problem.

Phase Four: Presenting and Implementing the Design This last step in the design process involves pulling all the pieces together so that you can present your design in a complete and compelling manner. It requires organizing and displaying the feasibility of your design in a way that helps others (namely, your client or, in an academic setting, your professor and fellow students) appreciate how it will meet their needs. Your design may represent a great idea or innovation but it will be of value only if you are able to convince others to use it. This is where presenting and implementing your design becomes important.

In 1857, Elisha Graves Otis demonstrated an innovative braking system on his passenger elevator at an exposition in New York City: he cut the elevator's cables while it ascended a 300-ft tower, and the brakes stopped the elevator from falling. His design became a huge success. You may never have to be this dramatic, but you will have to provide credible documentation that is easy to follow and you will have to be able to show that you have considered all relevant details in order to develop your design. In classroom settings, this aspect of the design process usually ends with the presentation and/or the final report. This text will give you some guidelines on how to do both effectively and will provide you with some insight to how design implementation occurs in industry.

1.2.3 Skills and Activities

As you progress through your engineering education you will be learning many essential technical concepts that apply to your chosen majors. Obviously, you cannot be an engineer without this discipline-specific knowledge. Of equal importance, however, are the four professional skill areas we will be discussing in this text (decision making, project management, communication, and collaboration). An understanding and appreciation of these skills will help to ensure that you can put your growing technical knowledge to its greatest use.

In fact, these four skills are so important to today's engineers that for a school to receive accreditation from the Accreditation Board for Engineering and Technology (ABET), it must demonstrate that its graduates have mastered them. For instance, according to ABET, engineering programs must demonstrate that their students have "an ability to function on multi-disciplinary teams." Collaboration, communication, and project management skills are essential to success in a team environment. ABET also requires that graduates have "an ability to identify, formulate, and solve engineering problems." Decision making is an inherent part of this ability. ABET also specifies that graduates must have "an ability to communicate effectively."

Like all skills, the best way to master the four skill categories we describe is through understanding, experience, and practice. These professional skills are usually subjects of separate textbooks. However, our goals are to introduce these skills to you in the context of engineering design and to provide you with some resources so that you can practice them and begin applying them to your own engineering work.

As you move through the various phases of the design process, different aspects of each skill become important or need to be emphasized. For example, different kinds of decision-making skills are required during the initial phases of a design problem than at the point when a solution is being proposed. At the initial stages, you need to understand specific skills and techniques for defining problems. Later, the focus shifts to being able to identify alternatives followed by an evaluation of the adequacy of those alternatives. In all instances, at each point in the process, you are making decisions, but there are different tools and procedures that you will want to master in order to make the best possible decisions. In a similar way, the various aspects of the other skill areas are presented at appropriate times. Thus, the materials for each of the four skill areas in this textbook are designed to move through the four phases of the design process with you. The design process consists of separate sections that will help you better understand each skill area as it relates to each phase of the process.

Table 1.1 provides an overview of how different tools and techniques for each skill area relate to the four phases of the design process. Each skill area is briefly described, and, although a number of the tools are useful at multiple points during the design process, we have tried to introduce them in an order that provides an understanding of the tool "just in time," borrowing a phrase from modern inventory control and manufacturing processes. By the time you complete this text and your design course, you will have had ample opportunity to understand, experience, and practice all the tools in each of the skill areas outlined in Table 1.1.

Decision Making As problem solvers, engineers are constantly making decisions. They need to be sure the decisions they make are sound and well-reasoned. They do this by systematically gathering information and evaluating alternatives. Decision making for engineers includes integrating diverse ideas, creating innovative solutions, and ensuring that a clear rationale forms the basis for each decision made. The kinds of decision-making activities you will learn include how to define a problem, eliminating biases, how to establish decision-making criteria, data gathering, brainstorming, and testing. You will also learn how these activities fit together to create a systematic and iterative approach to solving engineering problems.

Project Management Project management skills focus on the actual tasks and activities that need to be performed. Project management is the organizational form best suited for engineering, and its long history extends at least to the building of the great Egyptian pyramids. This means the challenge of managing projects has also been around a long time. In this text, we will describe some of the project management issues that must be considered and introduce you to some tools and techniques you can use for managing your own design projects. These include creating working agreements, establishing priorities, using Gantt charts, and record keeping. We insist upon Continuous Quality Improvement (CQI) throughout the project rather than quality inspected in at the end of the design process.

Communication Today's engineers must interact with many people from within and outside their profession. Good technical skills are essential to your success as an engineer, but they are of little use if you cannot clearly communicate your ideas and convey an understanding of client needs and perspectives. As one manager of engineers in a Fortune 500 company recently stated, "I don't think it's possible to go through the ranks now and not have good people skills and good communication skills."

Your ability to write and speak in a clear, engaging manner will help you convey your thoughts and win support for your ideas. At the same time, your abilities to listen well and to continuously probe for information can help ensure that you fully understand the problems you are solving. Communication also helps to create an environment in which the technical information, so critical to your success, will flow freely.

Collaboration As we mentioned at the beginning of this chapter, history is filled with heroic images of solitary inventors and engineers tinkering in their workshops to develop new ideas, innovations, and products, from Gutenberg's printing press to Benjamin Franklin's lightning rod to Thomas Edison's light bulb. The inquisitiveness, tenacity, and creativity of these individuals are traits to be admired and emulated, but the reality of our complex world is that engineering initiatives are almost always multifunctional efforts requiring the coordinated input of many people.

This kind of integrated effort characterizes the great engineering achievements of today. The Pathfinder mission to Mars, the personal computer, and the Internet are all examples of multifunctional engineering achievements based on teamwork and collaboration.

TABLE 1.1 Relationship of Skill Areas to the Phases of the Design Process

	Skills and tools for the process			
Steps in the engineering design process	**Decision making**	**Project management**	**Communication**	**Collaboration**
Phase one: Defining the problem 1. Forming the problem statement 2. Identifying functional requirements 3. Recognizing constraints and limitations 4. Defining a schedule and forming a team	• Research and data gathering • Eliminating biases and overcoming assumptions • Analyzing key phrases • Using objective trees • Using sketches • Clarifying the problem over time	• Discussing and defining project expectations • Coordinating schedules and planning meetings • Establishing working agreements • Adhering to your working agreement	• Active listening and probing skills • Laboratory record book • Composition skills	• Group formation and development
Phase two: Formulating solutions 1. Defining design parameters 2. Identifying alternatives 3. Evaluating and analyzing alternatives 4. Selecting a solution	• Innovation versus origination • Considering external factors • Brainstorming • Nominal group and delphi techniques • Lateral thinking • Systematic decision grids • Force field analyses • Making estimates	• Preparing and using Gantt charts • PERT/CMP techniques • Establishing and maintaining records	• Sharing the data gathered • Writing proposals • Preparing bibliographies	• Ensuring open participation • Reaching consensus and building commitment • Managing conflict • Avoiding groupthink

Phase three: Developing models and prototypes 1. Selecting a modeling process 2. Performing a design analysis 3. Testing the overall design 4. Revising, refining, and critiquing the design	• Quantitative and qualitative analysis • Conducting design and critical reviews	• Clarifying roles and responsibilities • Obtaining resources	• Writing progress reports • Providing feedback • Seeking input and feedback	• Managing role conflict and role ambiguity • Recognizing style differences • Eliminating social loafing
Phase four: Presenting and implementing the design 1. Presenting the final design 2. Implementing production 3. Introducing and distributing 4. Following up	• Dealing with last-minute changes, retrofits, and workarounds • Checklisting • Seeking a fresh perspective	• Assuring quality management • Applying Continous Quality Improvement (CQI) to your design project • Reviewing performance	• Developing presentation skills • Preparing visual displays • Making the presentation • Writing final reports	• Involving all team members • Reviewing team effectiveness • Celebrating success

Teams are a way of life for engineers. In the words of one engineering manager, "Each way you slice it, it's some sort of team structure that is making things happen, not just individuals." In this text, you will learn more about the collaboration skills required of productive team members. These skills include things like managing conflict constructively and ensuring open participation, as well as a general awareness of team development.

Synergies among the Skill Areas When students work on a team project of several weeks or a term (quarter or semester) they must apply all four skill areas in an integrated way. Students write a team working agreement that details how they will make decisions and work together (decision making, collaborating). They develop a schedule with a series of tasks, people assignments, a timeline (beginning and due dates), estimates of time with a column for actual time provided, and a set of columns that show percentage complete for each task (project management). They also have to plan when and where they are going to meet (project management, collaboration, communication). The team must establish how they are going to document what they do and when they are going to do it—meeting notes, analysis of components, progress reports, final reports, and oral presentations (project management, communication).

1.2.4 Case Examples, Projects, and Other Learning Tools

To help you learn about and apply the four skills described, the text provides a variety of case examples, projects, assessment tools, practice exercises, and review questions. Before you begin working with them we want to provide you with a quick overview of what they are and how they can help.

Case Examples Most of our examples are drawn from student projects with which we have been involved. Hopefully, you will see that these examples are not unlike your own design projects. Although the students working on these projects all struggled at times, they eventually succeeded by applying the principles, skills, and techniques about which you will read.

Dorm Room Design Project A project to design a dorm room is ongoing throughout the text. You will find different assignments for this project in each of the remaining chapters. We chose a dorm room, because we thought it was something most students, regardless of engineering discipline, could relate to. On the otherhand, you may already have an ongoing design project as part of your course. If this is the case, you can substitute your project for the dorm room and complete the assignments in a way that will help you finish your design project.

Assessment Tools You will also find assessment tools and other questions for reflection. For instance Chapter 5 contains a conflict management questionnaire that you can complete with your design team members. In addition, each chapter ends with a behavioral rating form you can use to self-assess your own and your design team's effectiveness.

Exercises and Review Questions Practice exercises have been woven into each chapter. These are brief assignments that will help you better understand a particular topic or skill. In addition, you will find review questions at the end of each chapter. Answering these questions provides you with an opportunity to summarize the main themes for a chapter.

1.2.5 Tips on Using this Text

We have written this text to be a resource tool that will support your own design experiences and there are at least two ways to use it. The first way is to move through the material sequentially. This approach enables you to learn about each phase of the design process at roughly the point at which you are actually working on that phase (just-in-time). In other words, read the chapters on defining a design problem (Chapters 2 and 3) just before or while you are working on defining your own design problem. Once you have more or less completed that phase of your design project, read the chapters on formulating design solutions (Chapters 4 and 5) and so on.

In general, you will find that the even-numbered chapters in this text introduce you to the specific steps associated with a design phase. They also describe the decision-making skills you will need to execute those steps. The odd-numbered chapters describe how project management, communication, and collaboration skills can be used in relation to each phase. For example, Chapter 2 explains the steps relating to Phase 1 of the design process—defining the problem. It also introduces related decision-making techniques like analyzing key phrases. Chapter 3 also deals with defining the problem; however, it focuses on particular project management, communication, and collaboration skills that are likely to help during Phase 1.

We strongly recommend the sequential, just-in-time approach if you are learning about engineering design for the first time. For a clear picture of how this approach organizes the text material, refer to Table 1.1 and look at each row of the matrix.

Alternatively, you may want to use the text more selectively if you have some prior experience with engineering design. Recall, the chapters discuss specific skill areas in relation to phases of the design process. However, perhaps your goal is to improve your project management skills (or decision-making, communication, or collaboration skills). If this is the case, you should focus on those sections that pertain specifically to the skill(s) that interests you. Look at each column of the matrix in Table 1.1 for a clear picture of the content highlighted by this selective, skill-focused approach.

The approach that works best for you depends on the extent of your prior design experiences and, of course, the guidance of your course instructor. In any event, we hope you find this text helpful during your current course or project and also as a resource that supports design projects you undertake in the future.

1.3 OTHER IMPORTANT CHARACTERISTICS OF THE DESIGN PROCESS

We have told you about our text but before you begin working with it, some additional background will be helpful. In this section, we briefly describe some key concepts that characterize engineering design. You will see most of these concepts reflected in your own design projects and some familiarity with them will better prepare you to deal with the challenges they pose.

1.3.1 Products and Systems

Engineering design consists of products and systems. Products are machines, components, or devices that people actually use. The system is the supporting environment that makes it possible to use the product. For example, a light bulb is a product but the receptacles for the light bulb—the wiring and power grid and the electrical power generating plant—constitute the system that makes the light bulb functional.

Similarly, the automobile is the engineering design product but the electrical and mechanical controls, the roadways, bridges, and traffic signals constitute its system. A team of civil, mechanical, electrical, and computer engineers are included in this transportation system design. Products and systems are highly interdependent. When designing a product it is critical that you understand the system in which it will operate or be a part of. By the same token, you cannot successfully design a system without comprehending the related products.

1.3.2 Multiple Goals

One of the biggest challenges designers face is how to create something that successfully satisfies a number of criteria, many of which may appear to be contradictory. For example, the book *All Corvettes Are Red* chronicles the reinvention of GM's legendary sports car in the early 1990s. The Corvette team set a number of design goals: The new model had to be as fast or faster than the model it was replacing, it had to meet the new crash standards, it had to meet the new fuel economy specifications, it still had to have traditional Corvette looks, and it could not cost more than the current model. Thus, as the team reached each stage of the design process, they had to find ways to address each goal. Typically, a very fast car is not very fuel efficient. However, if the car can be made lighter, it can be fuel efficient and fast. As the process evolved, the engineers developed a formula that said they could increase the production costs by $10 per car to save 1 kg of weight. Another innovative contribution to the lighter, stronger philosophy was the design of the side rails that formed the outer portion of the chassis. Closed, thin-walled tubes were filled with water and then high pressure was used to form the tubes in molds (hydroforming). This allowed the design to be strong enough to meet the side impact standards, and it saved weight over conventional manufacturing methods.

1.3.3 Closed Versus Open-Ended Problems

Engineering design is about solving problems, that range from simple one-answer problems (such as knowing the height a ball can reach given an initial velocity and direction) to more complex problems (such as determining the terminal velocity of a rocket with more than one booster). However, most problems engineers have to solve are open-ended and have several possible solutions. The challenge is to identify an optimal solution by careful and systematic analysis of multiple alternatives.

1.3.4 Application of Established Scientific Principles and Technologies

The information, materials, and processes that are available at a particular time will significantly impact the design solution that is reached and, for that matter, how a problem is conceived in the first place. An example from history illustrates this point. You may know that during the late 1400s and early 1500s Leonardo DaVinci attempted to design a flying machine. Fascinated by the movement of a bird's wings, he based his design on his observations of their graceful and effortless propulsion. His designs essentially imitated nature by equipping a person with wings to flap. However, it is a good thing he never built any of his designs because they would not have worked. Even in its simplest formulation, flying requires lifting a load from the ground against the field of gravity (lift) and moving it in a specified direction (thrust). Birds, with their remarkable power-to-weight ratio, can easily master these requirements. During Leonardo's time the concept of gravity had not even been defined let alone other key concepts such as lift and thrust! Had he been aware of them he might have recognized the futility of his designs. The natural materials of his day (wood, leather, cloth, metal) would have been too heavy. In fact, it is estimated that a person propelling a machine made from those materials would need the strength of 100 people. Today, however, with our knowledge of aerodynamics and the existence of high-tech lightweight materials, it is possible to design a human powered flying machine.[1]

1.3.5 Iteration and Feedback

At every stage in the design process something new will pop up requiring you to reexamine what you have already done and try something again, perhaps differently or repeatedly. As your team works together and gathers information on materials, devices, processes, and human factors, you will almost always find that the problem needs to be defined more clearly and that you need to repeat some steps by using new knowledge or perhaps applying a new technology. As shown in Figure 1.5, closer examination of your design may show that the materials you use and/or your arrangement of them simply will not allow the problem to be solved as the team initially perceived the solution.

Making these choices requires good decision-making skills. For example, in the book *All Corvettes Are Red*, the design team found that they could save weight by making the floor of the passenger compartment out of balsa wood instead of metal. The weight savings were needed so that the car could satisfy more stringent government regulations coming from the Corporate Average Fuel Economy (C.A.F.E.). Needless to say, this was an innovative solution that had not been anticipated.

[1]This example is adapted from *Discovering the Principles of Design Through Reverse Engineering* by Jean LeMee and John Razukas, Gateway Coalition, 1997.

FIGURE 1.5 Iteration is critical to design success. (Gary Larson/Universal Press Syndicate.)

During the product design process, team members must keep others informed about design decisions and changes. When particular components are tested or analyzed, it is critical for the rest of the team to be made aware of the problems. Sometimes pressure to get the job done, meet a deadline, or contain costs discourage iteration and feedback. In the long run this is almost always a mistake. Their importance is illustrated in a book titled *In the Name of Profit*. The book depicts conflict between ethics and desire for profit. One corporation was designing brakes for a new aircraft. The test engineers experimented with one brake design, found that it did not work, and immediately provided the information to the design team. However, the head of the design team did not want to admit that the design would not work. Communication just shut down. Iteration and feedback are critical to progress, and team members must be willing to listen even when disturbing information is provided.

1.3.6 Stepwise, Sequential Design vs. Concurrent Design

Through the early 1950s to the late 1970s and into the 1980s, design and subsequent production were typically viewed as a sequential process in which the manufacturing or construction teams did not become involved until the design engineering team completed all their work. This provided a structured way for engineers to approach

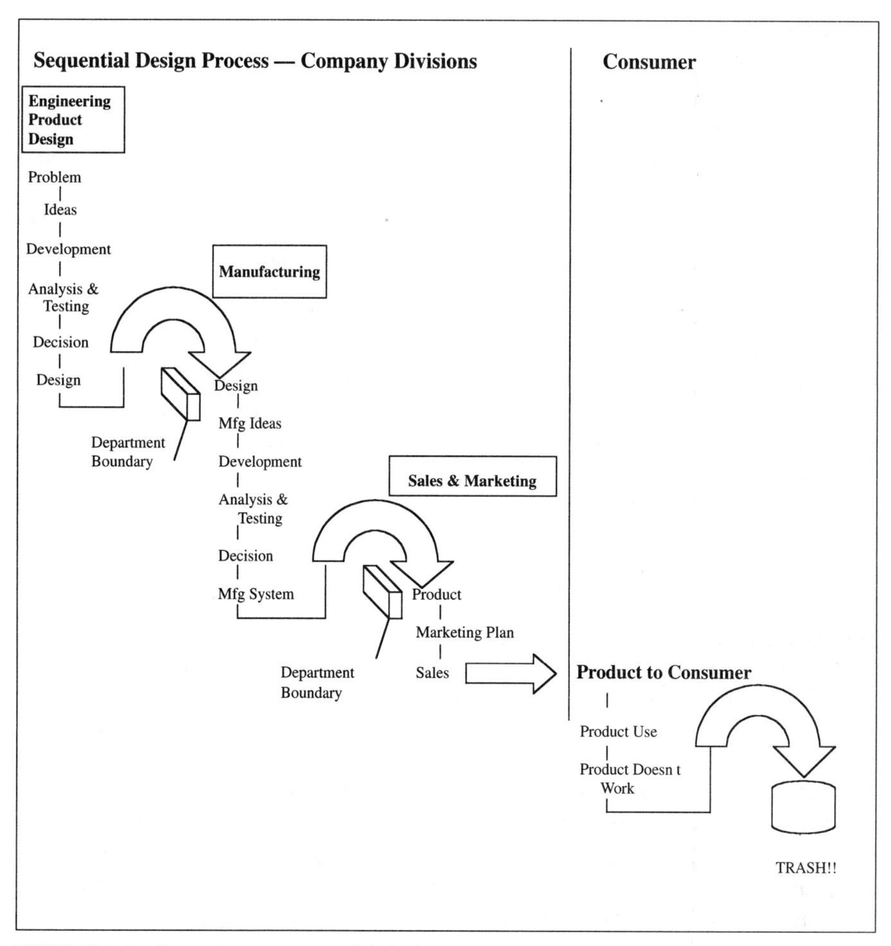

FIGURE 1.6 Stepwise or sequential design process.

a design or process that eliminated some problems and, as industry learned later, created others (Figure 1.6).

Many times a team believed they had the solution for a particular problem when, in fact, they really did not. As it turns out, this single-minded, single-track approach simply perpetuated an error of thinking that occurred early in the design process until the discovery was made at the end of the entire design process. The design team then had to do the ultimate iteration and go back to the start.

The main problem with the traditional design process is that a second design process began only after the initial design was completed. For example, when the manufacturing (construction) engineers got a preliminary design, they had to develop a manufacturing (construction) process to bring the design to life. As they looked at a design in light of the tools available, they found that parts could not be produced as originally conceived by the product designers. They then sent it back to the original design team or redesigned the part themselves. In other instances, the manufacturing engineers spent considerable effort and capital to manufacture a feature that was not necessarily of the greatest importance. Either approach meant that more time was required to get from the idea stage to the finished product. Communication with the design team early in the process could have resulted in substantial savings.

As companies find they need to bring products to the marketplace at an increasing rate and in shorter time periods, they have evolved the design process so that the initial design team includes those who have to produce, assemble, and maintain the design. In addition, a number of other areas of the company, such as marketing, finance, field support, and service, are represented on the design team. This is known as concurrent engineering. The design of the product is done in parallel with the design of the process used to manufacture the product and both are consistent with the needs of the marketing activities, field service, and technical support. This means good communication. Today design team members may even be physically separated, but the information, data, and decisions must be easily and efficiently shared.

Concurrent designing (Figure 1.7) requires thinking about many things simultaneously. For instance, in addition to the design and manufacturing considerations, thought must be given to the distribution, use, maintenance, operation, probable effect of a product or process on the environment, and eventual recycling not just at the end of production but throughout the design/build process.

When team members plan how they are going to bring about a new design, they must consider how much time it will take for each part of the process and which team members are going to be involved. They also have to figure out how much longer it will take to get into production and how much time will be required for resolving construction or manufacturing problems. These activities are all examples of project management skills and are dealt with in the following chapters. Before they start the process, the team must determine how they will work together and how they will resolve problems and conflicts (collaboration skills). These topics are also covered in later chapters in this text.

Although most college design projects are not going to be put into mass production or constructed, student teams can benefit from applying the concurrent design process. For example, if a prototype is to be produced or a model constructed, a successful team will begin experimenting with the production or assembly methods at a relatively early point in the project. This enables them to make sure they can complete the project on time and within budget.

Freshman design teams who build robots typically have constructed a chassis that is not "stiff" enough for the drive train so the chassis has to be redesigned. They also find that they have not done a good job of taking into account frictional losses in the drive train, and they have to generate new ideas for a solution that works. Smart teams do it concurrently.

Teams that participate in the American Society for Civil Engineers (ASCE) projects such as a steel bridge contest or a concrete canoe contest must learn about the materials and fastening methods early in the design process so they can do proper

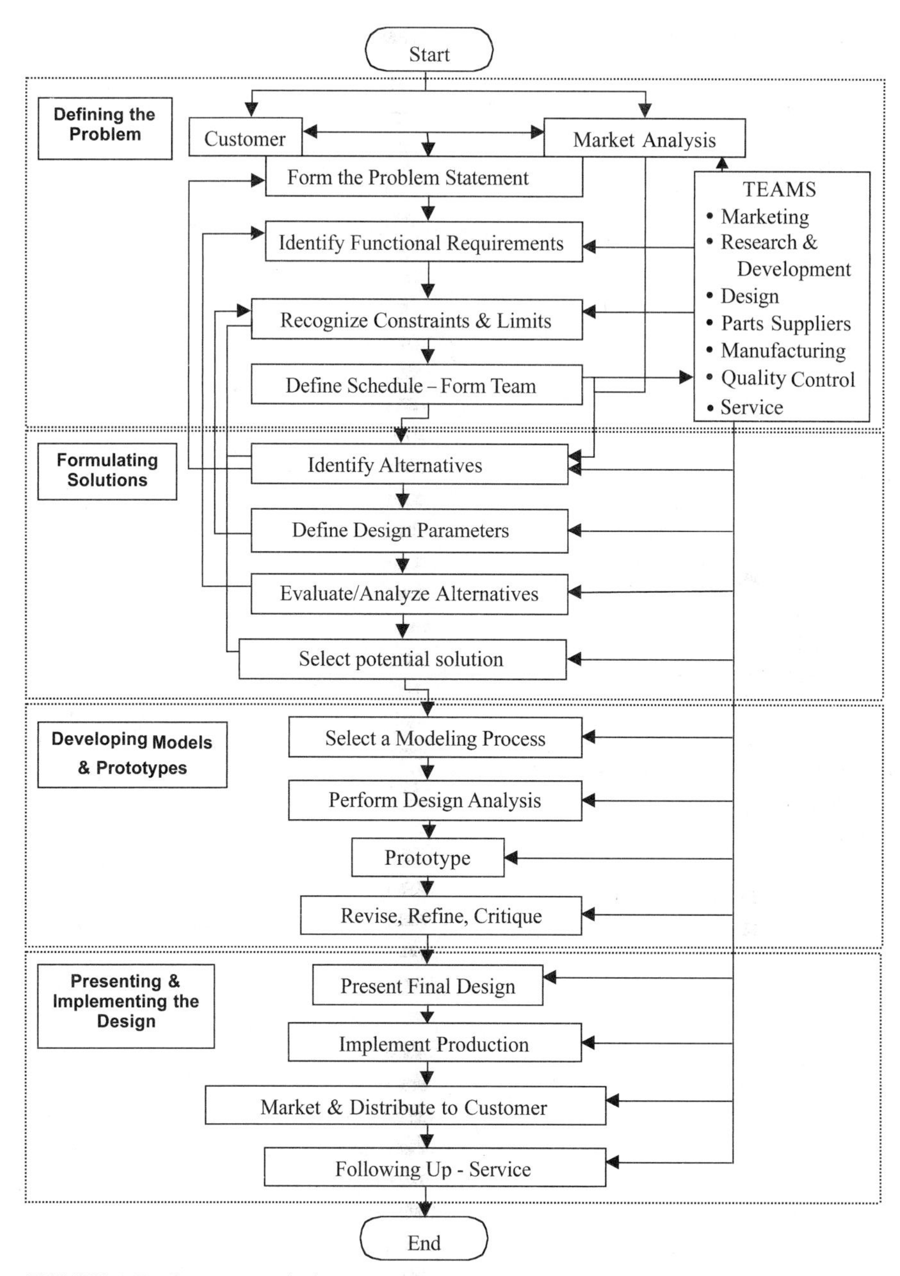

FIGURE 1.7 Concurrent design process.

design and analysis. Concurrent engineering helps greatly in this process to save time and rework effort.

The multischool design teams in the Gateway Engineering Education Coalition, described in the Preface, use a variety of methods to keep the geographically separate groups up to date. They have used e-mail, telephone conferences, videoconferences, and the World Wide Web to provide communications. They have evolved to a process whereby each school's group creates a Web site, and when they have a phone conference, all the Web pages are brought up to date and the design groups gather to discuss progress. As a group from one location talk, the other groups bring up their own Web pages to check their figures and calculations.

1.3.7 Safety and Reliability Issues

One of the many benefits of concurrent design processes includes increasing the likelihood that engineers can identify safety and reliability problems before a design is completed. These two issues must be addressed in every design, although they are more important in some instances than in others. We want to design all products so they will not harm or hurt others. This requires careful attention to details and the ability to visualize all ways the product might be used or misused. In the United States especially, society demands that engineers indentify and correct any product flaws that might hurt the consumer during the life of the product.

As the design evolves, the engineer should attempt to indentify any safety defects that might be present and correct them before they become a problem. For example, the assistive feeding device project that is discussed throughout this text involved designing a robot arm that attached to a wheelchair. During their work the student engineers recognized that potential users (quadriplegics) would have little fine motor control in their hands and arms. As a result, they would not necessarily be able to stop the arm from striking them in the face once the arm came close. The design team had to develop a control system that would restrict the range of motion of the robot arm. This was done by putting limit switches on the robot joints and programming the linkage to stop if the robot hand came too close to the user's face. Also, power to the joints was intentionally limited so the arm could be stalled if it contacted any part of the body.

As indicated before, the designer should look for any applicable design codes to apply to a given situation. These design codes often are developed with a viewpoint toward safety, and they should be used as a design guide even if the codes are not strictly required for the design situation.

In the "old days," engineers could concentrate on the intended function of a product, and they could assume that the consumer would be careful. If the consumer were careless or misused the product, the design engineer was unlikely to be considered responsible for any accidents. As a result, guards on machines and other products were rare, but amputations and other serious injuries were not! This is definitely not the case today.

Safety issues should be an integral part of the design process, and the design team should do a safety review as well as review that the product performs as intended.

In addition to checking the design relative to any applicable design rules or codes, the team should attempt to visualize how the product could be misused. Currently, society is placing more and more responsibility on the designer to anticipate how products can be misused to the detriment of the consumer. The team can use "brainstorming" (see Section 4.2.1) to indentify ways the product can be misused. They can then either design the product to remove unsafe situations or, at worst, place labels on the device to warm the user of dangerous conditions.

Reliability is similar to safety; however, the intent here is to produce a product that will perform its intended function for the design life of the part. Note that it is not necessary to design every part to last forever. In fact, designing for infinite life may be unnecessary and prohibitively expensive. The design team should determine how long the product must last as part of the design definition phase of the process. In practice, the design team is expected to warranty the part for its design life. If premature failure occurs, the designer's company is expected to repair or replace the defective part. To assess the reliability of product, the design team should perform a design review and assess the reliability of the product. The team must model the parts so the modes of failure can be identified. Typically, this involves functional analyses such as strength and heat transfer analyses. Ultimately, the designer would like to determine a factor of safety or the statistical chances of premature failure. Note that reliability issuse are more broad than safety issues. In safety issues, the designer is concerned about not harming the consumer, whereas reliability issues address preventing premature failure. Both should be addressed in the design stage before the product is manufactured. As in most aspects affecting product life cycles, the idea is to anticipate problems as early in the design cycle as possible. The longer a problem remains unresolved, the more expensive it is to resolve.

1.3.8 Ethical Issues in Design

Safety and reliability are closely related to another important design issue—ethics. You need to be aware that the technical challenges engineers face more often than not have broader implications for all aspects of society including economics, politics, privacy, and safety to name just a few. We cannot separate engineering and science from the broader social system in which it occurs. It is essential therefore that engineers have a constant awareness of the broader ramifications of their work.

They must also realize that their work often brings with it potentially conflicting obligations to a variety of stakeholders. These may include clients, consumers, workers, the government, and other engineering professionals. For example, one obligation engineers may have is to control costs and deliver a product on time. But what if, at the end of their work, they find out there is a slight chance their design will have unintended environmental consequences? Moreover, what if any further implementation delays would also cause many people to lose their jobs? Regardless of how this situation eventually might be resolved it is not too difficult to imagine how an engineer might feel pulled in different directions. No matter what our profession, we will be confronted with dilemmas that require us to make decisions based on our own internalized conceptions of right and wrong.

Our own internal standards are important, but what we mean by ethics is slightly different. The world is far too complex and diversified to expect that we all naturally share identical conceptions of right and wrong. Therefore, in terms of our roles as professional engineers, it is important that we rely on more universally defined guidelines and standards that can transcend our individual differences. Over the years such standards have been developed by engineering societies to help guide their members. Figure 1.8 shows the code of ethics adopted by the Institute of Electronics and Electrical Engineers (IEEE). Figure 1.9 lists another ethics code developed by the American Society for Civil Engineers (ASCE).

Although there are some differences between these two codes, both indicate how engineers should act with regard to clients, other engineers, and the general public. Moreover, the differences between them are a reflection of differences in the way electrical engineering and civil engineering are practiced. For instance, most civil engineers work in smaller firms. They frequently provide services to clients and obtain their work through public bidding. In contrast, electrical engineers frequently work for large corporations that sell products instead of services.[2]

Of course, the fact that standards exist does not mean that they, in and of themselves, resolve ethical dilemmas. These standards may help, but not all ethical situations are so black and white that we can just point to the standards for an answer. For instance, both the ASCE and IEEE codes stress that safety, health, and welfare should be paramount concerns. Later in this text, we describe how counterproductive group dynamics may have led to the launch decision that resulted in the explosion of the space shuttle Challenger. Those engineers who argued against the launch strongly believed that lives might be lost. It is difficult to argue, however, that the managers who favored launching did so because they did not care whether people would be hurt. Therefore, were these managers acting unethically? Perhaps one still could argue that they did not give enough ethical consideration to the potential (and, tragically, actual) consequences of their decision. But one also might argue that their decision, although clearly wrong in hindsight, was not made in an ethical vacuum.

The point of the above example is that ethical situations frequently are true dilemmas. We may not know for sure what the consequences of our actions will be. We can, however, do our best to acknowledge the broader responsibilities associated with our work and continue to develop our skills throughout our careers. We can also strive to always evaluate our work in relation to ethical standards and also to serve as sources of ethical support for one another.

CHAPTER REVIEW

In this chapter we have described how the text is organized and the recurrent themes that run through the subsequent chapters. Specifically, we have divided the design process into four distinct phases and have also indicated that there are four core skill areas for you to master. As you progress through each phase of the design process,

[2]C. Dym and P. Little. *Engineering Design: A Project-Based Introduction*, John Wiley & Sons, New York, 2000, p. 222.

IEEE CODE OF ETHICS

We the members of the IEEE, in recognition of the importance of our technologies in affecting the quality of life throughout the world, and in accepting a personal obligation to our profession, its members and the communities we serve, do hereby commit ourselves to the highest ethical and professional conduct and agree:

1. to accept responsibility in making engineering decisions consistent with the safety, health, and welfare of the public, and to disclose promptly factors that might endanger the public or the environment;
2. to avoid real or perceived conflicts of interest whenever possible, and to disclose them to affected parties when they do exist;
3. to be honest and realistic in stating claims or estimates based on available data;
4. to reject bribery in all its formats;
5. to improve the understanding of technology, its appropriate application, and potential consequences;
6. to maintain and improve our technical competence and to undertake technological tasks for others only if qualified by training or experience, or after full disclosure of pertinent limitations;
7. to seek and offer honest criticism of technical work, to acknowledge and correct errors, and to credit properly the contributions of others;
8. to treat fairly all persons regardless of such factors as race, religion, gender, disability, age, or national origin;
9. to avoid injuring others, their property, reputation, or employment by false or malicious action;
10. to assist colleagues and co-workers in their professional development and to support them in following this code of ethics.

FIGURE 1.8 IEEE code of ethics, August 1990.

ASCE CODE OF ETHICS

Fundamental Principles

Engineers uphold and advance the integrity, honor, and dignity of the engineering profession by:

1. using their knowledge and skill for the enhancement of human welfare and the environment;
2. being honest and impartial and serving with fidelity the public, their employers, and clients;
3. striving to increase the competence and prestige of the engineering profession; and
4. supporting the professional and technical societies of their disciplines.

Fundamental Canons

1. Engineers shall hold paramount the safety, health, and welfare of the public and shall strive to comply with the principles of sustainable development in the performance of their professional duties.
2. Engineers shall perform services only in areas of their competence.
3. Engineers shall issue public statements only in an objective and truthful manner.
4. Engineers shall act in professional matters for each employer or client as faithful agents or trustees and shall avoid conflicts of interest.
5. Engineers shall build their professional reputation on the merit of their services and shall not compete unfairly with others.
6. Engineers shall act in such a manner as to uphold and enhance the honor, integrity, and dignity of the engineering profession.
7. Engineers shall continue their professional development throughout their careers and shall provide opportunities for the professional development of those engineers under their supervision.

FIGURE 1.9 ASCE code of ethics, January 1977.

applying different aspects of each of these skills will help you get the job done. Keep in mind the matrix in Table 1.1 as you continue to read this text. It is essentially a road map to help you navigate your way through the material.

We have also described some key characteristics of all design processes. These characteristics further illustrate the relevance of the four design phases and the importance of developing your competence in each of the four skill areas.

REVIEW QUESTIONS

1. How has engineering changed over the years and in recent times? What implications do you think these changes have had on the skills required of engineers?

2. What is the difference between sequential and concurrent design? How do these differences impact design teams?

3. What makes engineering design different from pure science? In what ways do you think it might be the same?

4. What is the difference between a product and a system? From the perspective of engineering design, how are they distinct? How are they similar?

5. It is sometimes said that engineering design is as much an art as it is a science. Explain why you agree or disagree.

6. Why are teams an important aspect of engineering design? What do you think are some challenges teams pose?

7. What do we mean when we say that design is an iterative process?

8. How do ethical issues impact the engineering design process? Identify some of the major technological and scientific innovations of the past 10 years. For each that you identify, try to list some of the ethical issues they pose for engineers, scientists, and society in general.

9. Prepare a list of ethical guidelines appropriate for student design projects.

10. For each of the four skill areas (decision making, project management, communication, and collaboration) try to identify something you do well. Also identify something you would like to learn about or improve upon.

11. Explain what the terms safety and reliability mean. How are these concepts similar and how are they different? Why are they both important?

BIBLIOGRAPHY

The American Experience Technology Timeline. Public Broadcast System Website (www.pbs.org), 1998.

BERTOLINE, G., WIEBE, E., MILLER, C., and NASMAN, L. *Engineering Graphics Communication*. Irwin, Chicago, 1995.

BILLINGTON, D. P. *The Innovators: The Engineering Pioneers Who Made America Modern*. Wiley, New York, 1996.

DYM, C. L. *Engineering Design: A Synthesis of Views*. Cambridge University Press, New York, 1994.

EARLE, J. H. *Engineering Design Graphics*, 9th ed. Addison-Wesley, Reading, MA, 1999.

Engineering Accreditation Commission of the Accreditation Board for Engineering and Technology. *Engineering Criteria 2000*, 2nd ed. Accreditation Board for Engineering Technology, Inc., Baltimore, 1998.

HEILBRONER, R. L. *In the Name of Profit*. Doubleday Books, New York, 1972.

HUGHES, T. P. *Rescuing Prometheus*, Pantheon Books, New York, 1998.

LE MEE, J. *Designing Product and Success*. Paper presented at "Designing for Product Success," The Cooper Union, New York, 1990.

LEMEE, J., and RAZUKAS, J. *Discovering the Principles of Design Through Reverse Engineering*. Gateway Coalition, 1997.

MCGOURTY, J., TARSHIS, L., and DOMINICK, P. *Idea Generation and Innovation: A Behavioral Model Based Upon the Practices of Exemplary Companies*. Technical Report, Stevens Alliance for Technology Management, Hoboken, NJ, 1994.

SCHEFTER, J. *All Corvettes Are Red: Inside the Rebirth of an American Legend*. Pocket Books, New York, 1998.

VOLAND, G. *Engineering by Design*. Addison-Wesley, Reading, MA, 1999.

CHAPTER 2

DEFINING THE PROBLEM: STEPS AND DECISION-MAKING SKILLS

Clearly defining the problem at hand is the focus of the first phase in the engineering design process. It poses what very well may be the most critically important issues faced by you and your design team. Traditionally, many problems faced by engineering students have a single solution. But real-world engineering problems often are open-ended and not well defined. When presented with an open-ended problem, engineers must have a strategy for attacking and solving it. This phase of the engineering design process requires carefully executed steps to define the problem. Those steps are described in this chapter.

In executing those steps, you and your team will need to make a number of decisions and choices. Indeed, deciding and choosing are integral aspects of the steps throughout the entire design process. Informed, reasoned, unbiased, logical, and innovative choices are preferred here. In the early going, most decision making will be related to identifying the problem and clarifying or redefining it based on your research and on information gathered by team members. One of the challenges during this phase of the design process is to avoid introducing biases or prejudices that may preclude choosing a design alternative that otherwise would prove most successful. You also will need to avoid making choices and decisions too quickly. Over the years, engineers and other problem solvers have developed a variety of decision-making techniques that help meet these challenges. In this chapter, we introduce you to these tools, techniques, and tactics. Using them can increase the odds that you will ask the right questions and develop a more complete understanding of a problem.

Before reading on, briefly review Table 2.1. It provides an overview of Phase 1 by listing its main steps along with associated decision-making, project management, communication, and collaboration skills. The shaded boxes in Table 2.1 identify the topics covered in this chapter. The topics listed in the nonshaded boxes are discussed in Chapter 3.

For now, let us begin by introducing and defining the main steps related to defining a design problem. They include the following: forming the problem statement, identifying functional requirements, recognizing constraints and limitations, and defining a schedule and forming a team. As we explore the steps, you and your team will have the opportunity to be actively involved in trying out your new skills.

TABLE 2.1 Overview of Design Phase 1: Defining the Problem

Steps for defining the problem	Skills and tools for defining the problem			
	Decision making	Project management	Communication	Collaboration
1. Forming the problem statement 2. Identifying functional requirements 3. Recognizing constraints and limitations 4. Defining a schedule and forming a team	• Research and data gathering • Eliminating biases and overcoming assumptions • Analyzing key phrases • Using objective trees • Using sketches • Clarifying the problem over time	• Discussing and defining project expectations • Coordinating schedules and planning meetings • Establishing working agreements • Adhering to your working agreement	• Active listening and probing skills • Laboratory record book • Composition skills	• Group formation development

2.1 FORMING THE PROBLEM STATEMENT

A problem statement is a written description of the problem to be solved. It should be a succinct statement of the problem so that each member understands what the team is to accomplish. In your design course, your instructor may have given you a problem statement, or you may have been asked to write one yourself. Among the first things to understand about most engineering problem statements is that they often have the following characteristics:

- They are open-ended, which means there is typically more than one acceptable solution to the problem.
- They are loosely structured, which means solutions for design problems are not found simply by applying precise formulas in structured ways. In fact, engineers never operate in an environment where they are limited only by the laws of nature. Constraints and limitations abound.
- They are to be viewed in a systems context, which means that, in addition to the laws of nature, an engineering problem also must take into account the human environment in which the design will function. As a result, engineers always must endeavor to bridge between the "desires" of people and the realities of "nature." Because of these practical considerations, a properly defined engineering problem also takes into account availability of resources such as time and money and human factors such as personal preference.

- They should be accompanied by sketches or drawings as appropriate. A picture is worth a thousand words to a client, and your professor may require several.

As an example, we will look at a project to design an assistive feeding device (robot arm) to be used by persons with disabilities who are confined to a wheelchair. The project was completed by a team of senior engineering students from several universities. Clients explained what they needed and described some of the functions the arm would perform. This led directly to a problem statement as shown below in Figure 2.1.

Let us take a closer look at the problem statement in Figure 2.1. The heart of the statement is "A need exists for a simple feeding device." A rationale for this need is provided in the background information above the problem statement. We are also given an idea of what "simple" means because the preceding sentence declares that existing devices are "often bulky, expensive, and not portable." You should notice, however, that this problem is relatively open ended. No specific solution is implied; only some general criteria need to be satisfied (e.g., economical, flexible, low profile).

Problem Statement Background

Each year there are 7,800 to 10,000 spinal cord injuries in America resulting in a loss of body function. Of these injuries 51% are classified as quadriplegia which is defined as paralysis of the upper and lower body. Helping these individuals with simple daily routines dictates a need for assistive devices. The use of a feeding device would greatly improve their self-esteem and independence.

There are feeding devices available to help these individuals, but the devices have limitations. The users find existing devices to be bulky and not portable, and the devices do not have the flexibility to handle different kinds of food. The high cost of these devices limits the users to those that are financially well-off. Existing devices cost between $1,750.00 and $20,000.00. The lower priced devices are feeders only, and the higher priced devices are robotic arms that also assist individuals with other daily needs.

The Gateway Engineering Education Coalition, a group of researchers from 10 universities, has requested a feeding device be designed to assist people with disabilities. The design will be completed by engineering teams from Drexel University, The Ohio State University, University of Pennsylvania, and Cooper Union.

Problem Statement

Although feeding devices are available for quadriplegics they are often bulky, expensive, and not portable. A need exists for a simple feeding device. It should be electromechanical and operate from its own power source. The design should be economical, have a low profile, be flexible, and be simple to use and maintain.

FIGURE 2.1 Problem statement for the assistive robot arm design project.

In addition, the problem remains loosely structured. Nothing in the statement suggests that it can be solved by applying a precise formula or concept. This keeps the focus on defining the problem by not specifying a particular methodology for solving it. That will come later; to do so now limits possibilities and creativity. Finally, the statement gives us some idea of the broader system in which our design must function—namely, that it is to be used by quadriplegics. This context provides some general direction as we begin to develop this design problem further and we definitely want to know more about quadriplegia.

2.1.1 Research and Data Gathering

The way to find out more (about quadriplegia in the case of the assistive feeding device) is to conduct research and gather data. Making informed decisions requires understanding what currently exists or what has been done before that may relate directly to the problem at hand. Design teams will find it useful to gather data on related designs or products, physical features, current consumer or market trends, available technology, and the like. Traditionally, reference sources have included technical and trade magazines, new product announcements and specification sheets, patents, popular periodicals, and consultants among others. Fortunately, many of these traditional sources of information are now available in a useable form on the Internet with a suitable Web browser and search engine. Such ready access has dramatically reduced the time and effort required to conduct the research and gather data for better decision making. However, not everything on the Web has been tried, true, and tested. When reviewing a Web page consider the questions in Table 2.2. You should definitely be able to answer yes to the two questions in boldface type in order for the page to be considered credible. If you can answer yes to most of the other questions it is more likely that the source is high quality.

Part of the literature survey completed for the assistive robot arm design project is shown in Figure 2.2. In this case, students identified existing designs, determined their strengths and limitations, and ultimately used this information to determine how their own design could improve upon what was already developed.

Augmenting these traditional and new reference and data sources are opinion and informational surveys taken by personal interview, over the telephone, through the mail, or online. Student design teams also will likely find that the teaching faculty and the graduate and undergraduate teaching assistants are valuable sources of information. Surveys among potential users can help provide a better understanding of things like the market need for a product, its desired features, and acceptable price. For an example of the value of a survey among potential users, you can review the results for the robot arm project in Figure 2.3.

2.1.2 Eliminating Biases and Overcoming Assumptions

Certainly research and data gathering are important, but relying exclusively on research and existing data to clearly define a problem can sometimes impose a limitation that will impair a more in-depth understanding of your design problem. Specifically,

TABLE 2.2 Checklist for Informational Web Page

Web page criteria	Related questions
Authority	■ **Is it clear who is sponsoring the page?**
	■ Is there a link to the page describing the purpose of the sponsoring organization?
	■ **Is there a way of verifying the legitimacy of the page's sponsor (e.g., a phone number or postal address)? An e-mail address is not enough.**
	■ Is it clear who wrote the material and are the author's qualifications for writing on this topic clearly stated?
	■ If the material is protected by copyright, is the name of the copyright holder given?
Accuracy	■ Are the sources for any factual information clearly listed so they can be verified in another source?
	■ Is the information free of grammatical, spelling, and other typographical errors? (These kinds of errors not only indicate a lack of quality control but can actually produce inaccuracies in information.)
	■ Is it clear who has the ultimate responsibility for the accuracy of the content of the material?
	■ If there are charts and/or graphs containing statistical data, are they clearly labeled and easy to read?
Objectivity	■ Is the information provided as a public service?
	■ Is the information free of advertising?
	■ If there is any advertising on the page, is it clearly differentiated from the informational content?
Currency	■ Are the dates on the pages indicating when the page was written, when the page was first placed on the web, when the page was last revised?
	■ Are there any other indications that the page is kept current?
	■ If the material contains graphs and/or charts, is it clearly stated when the data were gathered?
	■ If the information is published in different editions, is it clearly labeled what edition the page is from?
Coverage	■ Is there an indication that the page has been completed and is not still under construction?
	■ If there is a print equivalent to the Web page, is there a clear indication of whether the entire work is available on the Web?
	■ If the material is from a work whose copyright has expired, has there been an effort made to update the material?

Source: Adapted from Jan Alexander and Marsha Ann Tate (1998). "Checklist for an Informational Web Page." Widener University, Chester, PA.

it is very easy for us to use the things that are familiar to us. These include common devices, ordinary systems and customary procedures. If a new problem can be solved or apparently solved by something that we know or have experience with, then we may very well be biased to use that procedure or device. Each of us may

Literature Survey

A literature survey on the existing designs was conducted prior to the start of the design process. Information was found at the library and the World Wide Web (WWW). Several promising designs were considered and their advantages and drawbacks were studied. This process was beneficial in the development of new ideas.

The first design that was explored was the Eatery feeding device developed by Maddak, Inc., shown below.

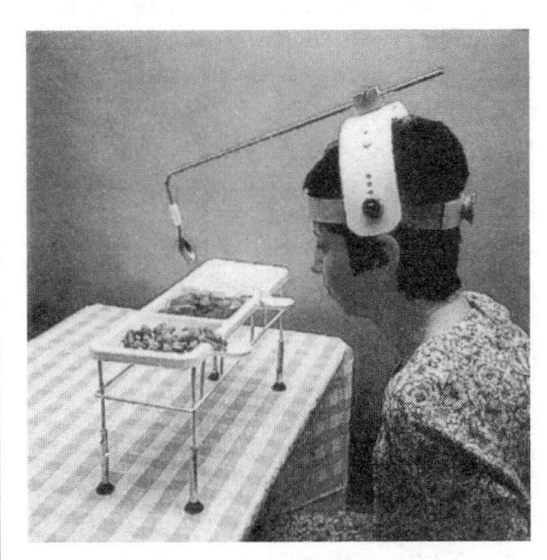

FIGURE—Eatery Feeding Device

The Eatery was composed of a height-adjustable stand with rubber feet, a plastic tray with three compartments, and a headpiece with a spoon attached to it. This commercially available device was designed for people without arm movements but with some degree of body and trunk movement since it required a lot of leaning motion in order to manipulate the spoon. The person would wear the headpiece and manipulate the spoon in such a way that the food could be placed on one of the two serving platforms. Then, the person could lean forward and take a bite.

The Eatery had more drawbacks than advantages because it was very difficult to use. It would be a tremendous challenge to use the spoon to perform the scooping motion, let alone placing the food on the serving platforms precisely. The food could easily be spilled and cleanliness would become a main concern. For to the reasons stated, the Eatery was not a desirable device because it could cause the user a great deal of frustration.

The second design considered was the One Step Mechanical Feeder that was developed by Mila, Inc. This feeding device was designed specially for people possessing leaning abilities and rough hand movements. The user could control the device by using a hand or an arm. The plate located on top of the base was indexed as the bar was pushed down. Therefore, it would rotate when the bar was pushed down and the food could be scooped up. This was a definite advantage over some of the designs considered because it would allow for the whole area of the plate to be reached by the spoon. A second benefit was simply that the spoon was brought up to the user's mouth and the user needs less range of motion to reach it. The attached cup holder was a good addition in theory, but if the patients are unable to manipulate a spoon to their mouths, it is unlikely that they could use a cup any more easily.

FIGURE 2.2 Literature survey results from the assistive robot arm project. (Courtesy Gary Kinzel.)

Synopsis of Survey Results

We used a structured interview to survey caretakers and potential end users in order to obtain their perspectives on the efficacy of the feeding device. Ten patients, seven caretakers, and a person working in rehabilitation robotics were interviewed individually or in small groups. The caretakers surveyed worked with quadriplegic patients and were responsible for feeding and assisting their patients. The same set of questions were asked of all interviewees and focused primarily on the perceived value of/interest in a feeding device and its functional requirements.

All interviewees responded positively to the goals of our project. Most patients indicated that they did not like the thought of having to rely on another person to care for them. All interviewees agreed that providing quadriplegics with the ability to feed themselves could enhance their self-esteem and potential for being more self-reliant. The kinds of requirements people said they would look for in such a device were that the device should be inexpensive, mobile, and flexible in all ways possible. They did not want it to be bulky and intimidating to them. When we asked how the device should be controlled, the general consensus was that it be somehow electronically controlled. Many of them suggested something like voice command or if possible, eye movement. One patient suggested that the device be manipulated with a mouth controlled joystick. Most others, however, felt a mouth-controlled approach would interfere with eating.

We learned additional useful information from a phone interview with an engineer specializing in rehabilitation robotics. He told us that most quadriplegic patients have mainly head movement and sometimes limited hand movement. He told us that there are feeding devices like the Handy 1 and another that consist of a rotating plate and a mouth-controlled utensil. He explained, however, that these current feeding devices handled only small morsels of food and could not accommodate liquids because they usually spill. Like one of the patients, he also recommended using a joystick-like device mounted at mouth level.

FIGURE 2.3 Synopsis of survey results from the assistive robot arm project.

bring a different bias or preconceived assumptions to the design process because of our different backgrounds in education, culture, experience, and personal preferences. The design process works best when biases and assumptions are recognized early. This does not mean, however, that we always need to "reinvent the wheel." The point is, by recognizing our biases and assumptions we are more likely to discover better, perhaps novel, or even revolutionary approaches to our problem. Consider this old problem. It may be familiar to many of you but it makes the point very well: *Without raising your pencil from the paper, link up the nine dots below using only four straight connected lines.*

● ● ●

● ● ●

● ● ●

When attempting to solve this problem many people assume that the lines cannot extend beyond the outer line of dots. Their preconceived notions compel them to "stay inside the box." They are biased by their tendency to see a square with borders. As a result, they impose limitations that make it impossible to solve the problem. Once they break out of this assumption, the problem becomes easy to solve, as shown below.

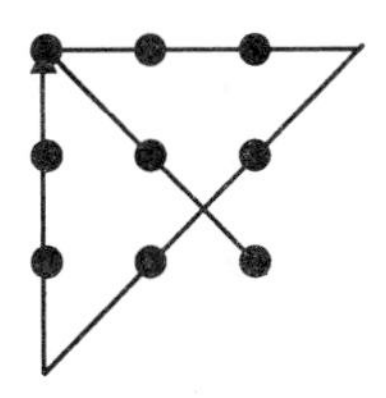

Now let's consider specifically how biases and assumptions relate to defining a design problem. In his book *Lateral Thinking: Creativity Step by Step*, Edward DeBono poses the following design problem:

> *Design a mechanical arm that will pick apples from a tree.*

To clearly identify this problem and develop a solution, DeBono suggests first considering the kinds of biases and assumptions that may be influencing our thinking. He recommends a very simple technique, repeatedly asking the question "Why?"

> *"Why does one need to* pick *apples off of the tree?" (They could be shaken off the tree instead.) "Why do apple trees have to be the shape they are?" (Perhaps the shape of the tree could be modified to make it easier to remove apples.) "Why does the arm have to go up and down with every apple it picks?" (The apple could be dropped into a chute or container.)*

In summary, assumptions and biases are inherent and often useful when it comes to looking at a problem. The point here is to understand where and when they exist so that we can also develop the skills to look past them and potentially obtain an even clearer picture of the design problem we are trying to solve.

2.1.3 Analyzing Key Phrases

In the previous example, simple words like "pick" and "arm" may have led many people to conceive of a design that did just that—use an arm-like structure to pick or grab apples from a tree. As we learned, however, there is usually more to a problem statement than what first meets the eye. Effective problem solvers start with the assumption that the initial problem statement is biased by a client's (and their own) limited perceptions and usually contains a preconceived notion of what the solution should be. It is crucial for the design team members to fully understand these implicit assumptions and biases. If they do not, they may fail to isolate the *real* problem. As a result, the design they create may fail to solve the underlying problem. In this section, we discuss another (related) technique for identifying and defining problems. It involves analyzing the key words and phrases in a problem statement.

Let's consider the following design problem that professor John Collier of Dartmouth University frequently presents to his new engineering students: *An ecologically concerned client has asked you to help her change her home heating system from electricity to gas in order to reduce her high heating costs.*[1]

If we proceed to solve our heating problem as is, we would focus on how to go about converting her heating system from electric to gas. But is this really the problem that needs to be solved? The statement is actually more of an implied solution than a problem. It reflects certain preconceived notions or biases on the part of the client:

- Heating with gas is less expensive than heating with electricity.
- Heating with electricity is both expensive and ecologically unsound.

Biases (as in this case) may contain elements of truth, but it is important for engineers to develop the discipline to dig deeper for their understanding of a problem. When we fail to make the distinction between implied solutions to the problem and the problem itself, we close ourselves off from alternative solutions that may be even better.

For instance, much thinking went into the mechanical design of various types of prototype tomato pickers before someone realized that the *real* problem was not coming up with the best design of a picker but rather in the bruising of tomatoes during picking. As a result, the solution became to develop a new tomato plant with tougher skin and more accessible fruit.

There are many reasons why people develop biases. Human beings are creatures of habit. We look for patterns to help us make sense of our world, but we can become so locked into these patterns that we fail to recognize other possibilities. One way to ensure that patterns are a help rather than a hindrance is to develop the ability to define problems in increasingly *specific* terms. This means looking for answers that help us clarify what the problem is and why it is occurring. It helps to ask questions like:

- What specifically is occurring?
- Who or what is being impacted by the situation to be addressed?
- Where or when is the problem occurring?
- How can we quantify the impact of the problem?

You may have noticed that these questions focus on defining the impact or effects of the problem. By asking these kinds of questions, we increase the likelihood of establishing a problem statement based on quantifiable information. It may seem ironic, but in order to establish a precise and specific problem statement we must first define the problem in the broader context in which it is occurring. This context includes not only technical issues but other factors such as economic conditions as well as political and social constraints.

Often a technically sound solution fails because the designer did not take into account these other factors (constraints) when initially defining the problem. Here is an example. Four freshman engineering students worked on designing an underground

[1]This sample problem is adapted from *Engineering Problem Solving for Mathematics, Science and Technology Education* by Ellen Frye (1997), Dartmouth Project for Teaching Engineering Problem Solving.

mail shuttle system for the elderly whose mailboxes were at the end of a long driveway. They identified the need for the product since many elderly people must hire someone to bring the mail to them or have family members become responsible for getting their mail. In their rush to solve this problem, however, the student engineers did not identify certain environmental factors that would impact the efficacy of their solution. They failed to isolate problems such as retrieving the underground mail if a mechanical failure occurred or problems with interference from power or water lines.

Let's see what happens when we start to isolate specific problems relating to our home-heating scenario. We can do this by systematically asking the kinds of questions described above. When we ask specifically what is occurring, we learn that our client is paying more money than necessary to heat her home. We also learn that she has records of her heating bills going back to when she first moved into her house. This will give us some idea of the impact of the problem. We also learn that the client believes her situation is ecologically unsound because higher costs also mean that more energy is being used.

By seeking quantifiable clarification of the original problem statement, we learn that there are actually two problems:

- The cost of heating her house is too high.
- Heating with electricity is both expensive and ecologically unsound.

Even these problems, however, can be broken down further. We can do this by focusing on some of the key phrases and notions embodied in the original problem statement.

- *Ecologically unsound:* How great is her concern? How much is she willing to pay for addressing this concern?
- *Change from electricity to gas:* Is gas always cheaper than electricity? How much does "change" cost?
- *High heating costs:* How do her heating costs compare with comparable homes in the area?

Our exploration of this problem eventually leads to a redefinition of the original problem statement. The new problem statement more specifically addresses her issues and concerns. It reads: *Identify the most cost-effective change the client can make to reduce her energy cost.*

When confronting a new problem statement, be prepared to analyze and ask questions about every key word or phrase in that statement. In particular, watch for phrases like "because of" or "due to." They generally imply a direct cause and effect and can imply a solution prematurely. For instance, consider the following statement: *The problem is that downtown stores are losing business because of poor traffic flow and parking limitations.* If we take this statement at face value, we begin to focus on changing traffic flow and parking space. In reality, this may or may not be the problem that needs to be solved. What questions can you ask to help clarify or redefine this problem statement?

2.2 IDENTIFYING FUNCTIONAL REQUIREMENTS

No matter the field of application, design is a process of interaction among a number of factors. First, there must be a need—a desire by someone somewhere (a client, a sponsor, a professor, a user, or a customer—for something the design process can bring forth. If no one wanted homes, builders soon would be out of work. If we did not need or want a means of personal transportation, the automotive industry would change dramatically. Second, there must be an agent who can understand these needs, meet them, and satisfy them. If there are no architects, no amount of wishing will make the office building appear. If none of us wanted a high performance, U.S.-made sports car, the Chevrolet Corvette probably would not be available. When both the agent and the client come together, through dialogue (communication), they establish the basic specifications for the design. These are the functional requirements. Functional requirements are the "what" of a design. What must it do? Clarifying these functional requirements is key to defining a problem in such a way that it can eventually lead to a useful design. The "how" of the design process can be arrived at only after the functional requirements are clarified.

Take a look at the example of functional requirements specified for the robot arm design project shown in Figure 2.4.

Of course, both agents and recipients have their own viewpoints of the problem to be solved. Their different backgrounds, training, interest in the project, and many other factors determine their viewpoints. Furthermore, these viewpoints may change as levels of understanding and awareness change. They are also impacted by the way

Functional Requirements

Size and Weight: The size and weight of the device should be minimized for the purpose of keeping the portability and aesthetic aspects of the design in mind. The materials used to construct this device should be chosen to meet this design requirement.

System Control: The control of the device should be simple to use, require minimal physical effort, yet still give the user complete control of which food to pick up and the rate at which he or she will eat.

Food Size, Weight, and Location: The feeding device will lift food items from a dinner plate located in front of the user. The size and shape of the food should be able to fit on the utensil; however, an ideal device would be able to handle any food size, weight, or texture.

Speed and Exertion Force: The speed at which the device feeds the individual should be relatively slow so as not to pose any physical hazards. The individual should also be able to control the rate at which he or she is eating.

Care and Maintenance: The feeding device must be easy to maintain, clean, and store during idle periods. Most importantly, it must provide the user with comfortable and lasting operation.

FIGURE 2.4 Functional requirements for the assistive robot arm or feeding device design project.

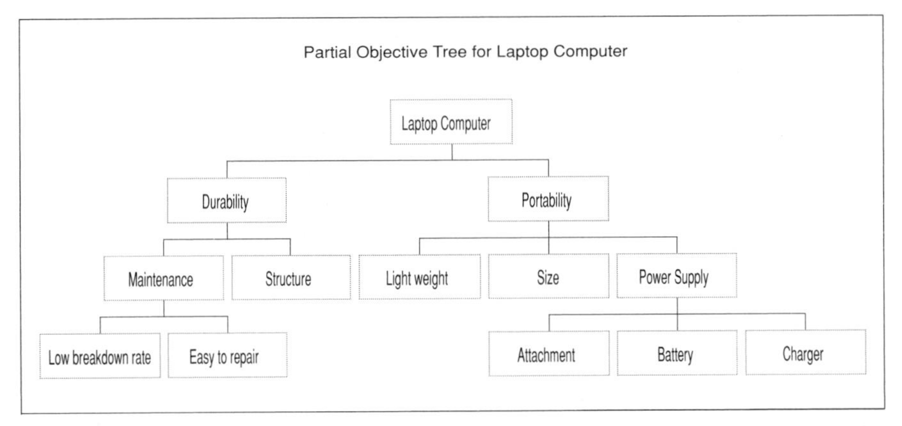

FIGURE 2.5 Objective tree for designing a laptop computer (adapted from Dym 1994).

people perceive their broader environments (e.g., their families, companies, communities). The interchange between the agent and the client provides the common ground that eventually results in an agreed upon set of specifications for the design.

2.2.1 Using Objective Trees

Hopefully, you now recognize that successful problem definition involves breaking a problem down into smaller components. Objective trees are another technique for identifying subcomponents of a design problem and can be particularly useful when trying to identify the specific functional requirements of your design. To create an objective tree you work downward, starting with the most abstract concept (usually the general problem statement) and then continually attempt to clarify what the concept means by decomposing it into more and more specific terms. Shown in Figure 2.5 is an example of an objective tree for designing a laptop computer. The initial problem posed to these designers was to "design a durable, low-cost, laptop computer."

Figure 2.5 is only a partial objective tree. This is because there are many more subcomponents to identify, such as modular design and replacement parts. In addition, most of those that are listed can also be broken down further. Take a moment to build upon this objective tree.

- Can you think of some other subcomponents to this design?
- What questions might you ask to break down some of the subcomponents that are listed in the diagram in Figure 2.5?

For instance, another component might be ergonomic characteristics, which in turn could be broken down into the keyboard and computer screen. As you might imagine, even these areas could be broken down further.

There are also some questions to ask that would help to further define components listed in Figure 2.5. How light do we want it? What are the size dimensions we want to achieve? What do we mean by "low breakdown rate"? Even if we do not yet have the answers to these questions, we are better off for having identified them systematically. The objective tree provides a kind of road map that can help you and your team identify specific tasks and activities that must be accomplished. Try to develop an objective tree in as much detail as possible, but do not be discouraged if it generates more questions than answers. You may be able to answer some questions through research and data gathering, although the answers to others may not yet be apparent. Therefore, like all aspects of design, expect to refine and build upon an objective tree as your design evolves over time.

2.3 RECOGNIZING CONSTRAINTS AND LIMITATIONS

At about the same time a problem statement begins to be formed and functional requirements are identified, the design process is immediately impacted by constraints and limitations that arise from a wide variety of sources. The "what" and "how" of a design are shaped in part by the "but" and "however" of constraints and limitations.

In the case of the assistive feeding device several factors posed constraints and limitations on the size of the arm when in use and when not in use. The addition of any new mass to a wheelchair can affect its handling and balance. The robot arm design teams had to determine where the center of gravity was located; where the existing wheelchair controls, power source, power transmission, and steering were located; whether the control for the arm would be added alongside the existing controls or integrated into the existing controls; what type of controller would be used; what sort of weight would be added; and so on. These considerations all posed practical constraints and limitations on the design. And certainly, the design process will always be constrained and limited by considerations based on time and money. An example of these types of constraints and limitations is given in Figure 2.6.

However, engineers often can be constrained in other ways because of a lack of experience or overreliance on previous work. It may seem strange, but original or novel ideas are sometimes not seriously considered and even may be discarded because they are just that, original or novel. Be open to new ideas.

The following constraints and criteria will impact the design:

- Users will have no hard arm movement.
- The design will take into account that the user does not have the anility to lean forward and is seated in wheelchair leaning back at an angle of 15 degrees.
- The speed of the device is constrained by the rate at which quadriplegics are able to safely chew and swallow food and as a result should not exceed one foot per second.
- The weight of the device is constrained by the fact that it must be portable.
- Design resources are limited and as a result we must use as many off-the-shelf products as possible.
- The cost cannot exceed $1500 for the final product.

FIGURE 2.6 Parameters, constraints, and criteria for the robot arm feeding device to assist people with disabilities.

2.3.1 Using Sketches

An important part of the problem definition (and the entire design process for that matter) is to make sketches of what you currently understand the problem (and system) to be. Often, drawing pictures of problems can help you see a problem in a new light and may help you to identify constraints and limitations that were not apparent when the problem was just defined in words. Numerous software programs allow you to do sketches electronically but sketches can be done on paper, in particular engineering problem paper. This paper is typically green three-ring paper with a grid on the back and spaces for project name, team name or number, date, and page. The person doing the sketch should be identified by writing "drawn by" at the bottom of the page. The sketches done to scale when possible and the scale should be identified (e.g., 1 unit = 6 cm). If it is not to scale, then it should be shown as "Scale: NTS." Figure 2.7 presents an example of one student team's design sketch for a robot.

2.3.2 Clarifying the Problem Over Time

As we mentioned in Chapter 1, design is an iterative process during which new information is likely to be uncovered. This point is particularly relevant when it comes to identifying constraints and limitations. At any number of points, sooner or later in the design process a design team may find that the problem statement simply does

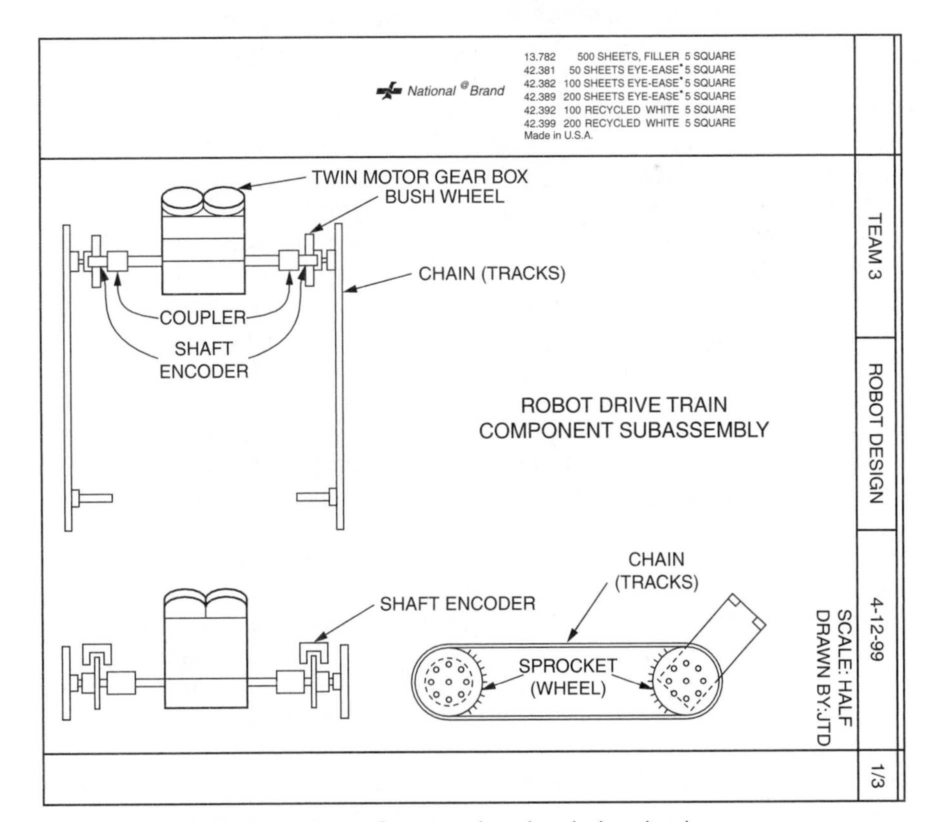

FIGURE 2.7 Typical page layout for an engineering design sketch.

not adequately describe all or enough of the aspects of the problem and its associated constraints. At such junctures, the statement of the problem needs to be rewritten, and the design must proceed back through all the previous steps of the process to ensure that new knowledge or new information is being incorporated where necessary. It is important to recognize that when a difficulty with the definition of the problem arises, the overall problem may be incorrectly stated or just some smaller portion may need revision.

In the autonomous robot design project, some student design teams discovered that the statement of the problem (or, more explicitly, the description of the terrain or course over which the robot must operate) implies that center of gravity considerations may be important. Some design teams locate heavy components on the robot chassis without enough consideration of the effect on the center of gravity. They typically do not investigate all the possible situations on the course where the robot is required to go up or down a slope or along an incline. When the robot tips over, the statement of the problem needs to be revised to include an understanding of the terrain over which the robot must operate (an important constraint).

2.4 DEFINING A SCHEDULE AND FORMING A TEAM

The three steps described in this chapter focused on the design problem itself. The fourth equally important step during this initial design phase involves planning and organizing how you will do the rest of the work. Typically, a schedule of the activities that must be performed in achieving a solution is prepared shortly after the problem is identified. At the same time, this is also the point at which you start transforming yourselves from a collection of individuals into a coordinated team.

There are many reasons and emphases to the process of forming a team. The initial emphasis for student engineering teams is on getting to know one another, establishing regular meeting times and locations, and gaining an understanding of the importance of coordinating personal schedules. Later you will learn about the functional and operational aspects of team formation and their attributes. Regardless of the actual nature of your design project, the significance of team formation and defining a schedule cannot be understated. Typical things that must be decided early on include identifying critical components and tasks and then sorting them according to what needs to be done first, what things can be done concurrently, and what logically comes later. A schedule of ordered activities and progress review deadlines may be provided by your course instructor. If not, you need to think about establishing your own schedule; in any event all team members must be fully aware of deadlines and their roles in meeting them. Student teams also must develop an overall schedule, outlining which team member will be working on each component or task, which member will serve as a back-up, and how much time should be set aside as a "cushion" against difficulties that may occur before a progress review deadline. In addition, your team should lay out some ground rules by establishing codes of conduct or agreements and describing what you expect from one another and how you will work together. An example of a schedule developed by students working on the autonomous robot design project at Ohio State University is shown in Figure 2.8.

ROBOTEAM										
TASK	Responsibility	Start Date	Target Date	Due Date	Time Est.	Actual Time	Percent Complete			
Develop Initital Design Plans	Jane, Ron	3/29/99	3/31/99	4/2/99	2 hours	3 hours	0	25	50	100
Develop Final Design Plans	Barbara, John	3/31/99	4/2/99	4/5/99	2 hours	3 hours	0	25	50	100
Develop Team Name	Jane, Barbara	3/31/99	4/2/99	4/16/99	30 min	1 hour	0	25	50	100
Buy Team Notebook	John, Ron	3/31/99	4/1/99	4/2/99	15 min	15 min	0	25	50	100
Create Team Contract	Jane, Ron	3/29/99	3/30/99	3/31/99	30 min	30 min	0	25	50	100
Finish Chassis Design	Barbara, Jane	4/4/99	4/7/99	4/9/99	1 hour	1.5 hr	0	25	50	100
Build Chassis Model	Jane, John	4/5/99	4/8/99	4/9/99	1 hour	2 hours	0	25	50	100
Design PVC Components	John, Barbara	4/7/99	4/7/99	4/9/99	10 min	10 min	0	25	50	100
Create Design Schedule	John, Jane	4/7/99	4/9/99	4/12/99	30 min	1.5 hrs	0	25	50	100
Decide Motor Configuration	Barbara, Ron	4/9/99	4/11/99	4/19/99	30 min	1 hour	0	25	50	100
Complete Chassis Design	Ron, John	4/11/99	4/12/99	4/19/99	15 min	45 min	0	25	50	100
Design Motor Mounts	Jane, Barbara	4/11/99	4/12/99	4/13/99	30 min	30 min	0	25	50	100
Buy Motor and Materials	Ron, Jane	4/12/99	4/12/99	4/14/99	10 min	10 min	0	25	50	100
Build Motor	Barbara, John	4/12/99	4/12/99	4/14/99	30 min	45 min	0	25	50	100
Develop Tread Design	Jane, John	4/10/99	4/12/99	4/14/99	45 min	30 min	0	25	50	100
Build Chassis	Ron, John	4/12/99	4/14/99	4/16/99	1.5 hrs	3.5 hrs	0	25	50	100
Performance Report 1	Barbara, Jane	4/14/99	4/16/99	4/19/99	45 min		0	25	50	100
Build Motor Mounts	Jane, Ron	4/15/99	4/19/99	4/23/99	1.5 hrs		0	25	50	100
Mount Motor on Chassis	Barbara, John		4/19/99	4/23/99	30 min		0	25	50	100
Attach Treads	Jane, John		4/19/99	4/23/99	1 hour		0	25	50	100
Complete Robot Base	Barbara, Jane		4/21/99	4/23/99	30 min		0	25	50	100
Develop Locomotion Code	John, Ron		4/21/99	4/23/99	30 min		0	25	50	100
Design Sensors Plan/Setup	John, Barbara		4/23/99	4/30/99	45 min		0	25	50	100
Complete Plan for Code	Jane, Ron		4/23/99	4/26/99	45 min		0	25	50	100
Performance Report 2	Barbara, Jane		4/21/99	4/23/99	45 min		0	25	50	100

FIGURE 2.8 Portions of a design schedule for the autonomous robot project at Ohio State University.

Building a schedule and forming your team are crucial for defining the design problem and avoiding unnecessary conflict. The engineering design team must function like an efficient machine to identify functional requirements, constraints, and limitations. On a very basic level, people must feel they can count on one another. Meetings of the full team need to be well coordinated and work must be spread out evenly. Project management, communication, and collaboration techniques for dealing with these important challenges are among the topics discussed in Chapter 3.

CHAPTER REVIEW

Phase 1 (defining the problem) is all about creating a foundation for the problem to be solved and the work to be done. You have learned that there is more to defining a problem than just reiterating the problem presented by your instructor (or your client). To put it succinctly, the answer you get (your design solution) depends on the questions you ask (your definition of the problem). So before we can identify solutions we must first break a problem into its component pieces by identifying specific functional requirements and then recognizing the constraints and limitations that surround them. A variety of decision-making tools and techniques can help you break down your problem and thereby obtain a more complete picture. They include doing thorough research, challenging your biases and assumptions, analyzing words and phrases, building objective trees, and even making sketches. These techniques have in common the facts that they promote systematic thinking, encourage you to look past superficial descriptions, and stress focusing on details. Use these factor to eventually get closer to understanding the root causes and issues of your engineering design problem. To help you review this chapter, answer the questions below.

REVIEW QUESTIONS

1. What is an engineering problem statement? Describe some of its important characteristics.

2. Section 2.2 begins by stating that "design is a process of interaction," Explain what is meant by that phrase.

3. What are functional requirements of a design? How are they different from constraints and limitations? In what ways do these two concepts complement one another?

4. What does it mean to *redefine* a design problem? What are the advantages of doing so systematically? Can you think of any disadvantages? What are some things you might do to redefine a problem?

5. Create an objective tree to help you better define requirements for the following design problem: *Design a cup holder system that fits most cars. It should hold a wide range of containers securely, be easily reachable to the driver, minimize spills and be easy to clean.*

What questions have you answered from creating this objective tree? What are some additional questions you have identified from building the objective tree?

6. How do biases and assumptions relate to defining a design problem? What are some things you can do to make sure they do not prevent you from fully defining a design problem? Can you think of any ways that biases and assumptions may help you define a problem?

7. How does the technique of analyzing key phrases relate to the concept of overcoming biases and assumptions?

8. Explain how a simple technique like drawing a sketch might help to define a design problem.

DORM ROOM DESIGN PROBLEMS FOR CHAPTER 2: DEFINING THE PROBLEM (STEPS AND DECISION-MAKING SKILLS)

Now that you have some understanding of the design steps and decision-making skills that relate to Phase 1, here is a problem for you to work on. The initial problem statement and first two assignments are described below. You will see other assignments relating to the same problem at the end of each chapter in this text.

Problem statement:

Warp University has surveyed the students living in dorms and has found that they would prefer to have individual rooms and share a bathroom rather than sharing a room with a bathroom. The job of your design team is to design the furniture necessary for a dorm room for one student. The door into the room swings inward and the door from the room to the shared bathroom swings into the bathroom. The university is willing to have sets of unique furniture components manufactured for this purpose.

Assignment 1 (Decision Making Skills)

Review the problem statement and determine if there is additional information that you need to know to continue the process. Your instructor, the library, and the Internet are your sources for additional information. Re-write the problem statement based on your new information. Make a sketch of the dorm room to scale and include it with your problem statement.

Assignment 2 (Decision Making Skills)

As a team of three or four members, review the furniture that the dorm room needs. Keep in mind all of the activities (studying, reading, sleeping, etc.) that go on in student rooms. Type out the list of furniture that you believe is needed. Include sizes. Write a brief report of your findings and include the list of furniture and sizes. Include sketches of the items to scale.

BIBLIOGRAPHY

ALEXANDER, J. E., and TATE, M. A. *Web Wisdom: How to Evaluate and Create Information on the Web.* Lawrence Erlbaum Associates, Mahwah, NH, 1999.

DEBONO, E. *Lateral Thinking: Creativity Step by Step.* Harper-Perennial, New York, 1990.

DYM, C. L. *Engineering Design: A Synthesis of Views*. Cambridge University Press, New York, 1994.

FENTIMAN, A. *Team Design Projects for Beginning Engineering Student*. Technical Report ETM-10-05-958. Gateway Engineering Education Coalition, Philadelphia, 1997.

FRYE, E. *Engineering Problem Solving for Mathematics, Science, and Technology Education*, Dartmouth Project for Teaching Engineering Problem Solving, Hanover, NH, 1997.

WEGGEL, R., ARMS, V., MAKUFKA, M., and MITCHELL, J. *Engineering Design for Freshmen*. Technical Report ETM-1058 Gateway Engineering Education Coalition, Philadelphia, 1998.

CHAPTER 3

DEFINING THE PROBLEM: PROJECT AND PEOPLE SKILLS

Now that you have an appreciation for the technical challenges posed by the first phase of the design process we want to focus next on dealing with some of the organizational and interpersonal challenges that are typically associated with the early stages of a design project. This chapter introduces you to specific tools and tactics pertaining to project management, communication, and collaboration. The topics covered are shown within the shaded boxes in Table 3.1.

You will be able to apply these skills at other points in your design work, but we think they are particularly critical during the defining the problem phase. For instance, project management at this point in the design process focuses on establishing the initial framework for how you and your team will complete your task, and you will learn about some specific actions that will create this necessary structure through meeting plans and working agreements. In terms of communication skills, Phase 1 of the design process emphasizes obtaining information and facts that provide a clear understanding of the problem. This means communication techniques like active listening and probing are most critical. Phase 1 of the design process is also the point at which most design teams are beginning to come together as a unit. Therefore, you need to apply some specific collaboration skills that will enable your team to develop into a cohesive and productive unit.

Table 3.2 relates specific skills to each design step discussed in the previous section. It provides an overview of when you will be most likely to benefit from using each skill and tool that is described.

3.1 PROJECT MANAGEMENT

Project management at this initial phase in the design process involves laying the groundwork for how you and your team will accomplish your work. In this section we provide you with guidelines for addressing three important topics to help orient your team in the correct direction for sure success. First, you and your team members need to reach an agreement about what each person must do and what you all expect in terms of results. Second, we will offer some suggestions about how you and your team can best coordinate your schedules and make certain that your meetings are well planned and productive. Finally, we will walk you through the steps you should

TABLE 3.1 Overview of Design Phase 1: Defining the Problem

Steps for defining the problem	Skills and tools for defining the problem			
	Decision making	**Project management**	**Communication**	**Collaboration**
1. Forming the problem statement 2. Identifying functional requirements 3. Recognizing constraints and limitations 4. Defining a schedule and forming a team	• Research and data gathering • Eliminating biases and overcoming assumptions • Analyzing key phrases • Using objective trees • Using sketches • Clarifying the problem over time	• Discussing and defining project expectations • Coordinating schedules and planning meetings • Establishing working agreements • Adhering to your working agreement	• Active listening and probing skills • Laboratory record book • Composition skills	• Group formation and development

TABLE 3.2 Relationship of Skills and Tools to Each Step in Phase 1 of the Design Process: Defining the Problem

	Phase 1 design steps			
Skills and tools	**Forming the problem statement**	**Identifying functional requirements**	**Recognizing constraints and limitations**	**Defining a schedule and forming a team**
Project management				
▪ Discussing and defining project expectations				X
▪ Coordinating schedules and planning meetings				X
▪ Establishing working agreements				X
▪ Adhering to your working agreement				X
Communication				
▪ Active listening and probing skills	X	X	X	X
▪ Laboratory record book	X	X	X	X
▪ Composition skills	X	X	X	X
Collaboration				
▪ Group formation and development				X

follow for creating a formal team agreement, describing what you expect from one another in the course of working together.

3.1.1 Discussing and Defining Project Expectations

One of the keys to the future success of any team is to make sure all team members share a common vision of what they hope to accomplish. You may be thinking that this vision is obvious: "We want to complete the assignments outlined in our course syllabus, meet the deadlines indicated by our instructor and get a good grade." This is indeed a good starting point for defining expectations, but you are making a big mistake if you stopped there. Do not take it for granted that you and your team members share identical interpretations of the course outline. Take time to review assignments as a group. Confirm that you all understand what work you will be doing individually and together, and try to identify any questions you have about your personal responsibilities. Perhaps you will be able to answer each other's questions about requirements and assessment. If necessary, seek the input of other students and, of course, your instructor. The point of these discussions is not only to get the facts straight (that is important) but also to begin building a collective understanding as a team of what it is you must do between now and the end of the term.

You should also use this time to learn more about each other's prior experience, interests, and areas of expertise. Perhaps some members of your team have had prior design experience or experience working on group projects. Discuss what these experiences were like. What did people learn about them? What challenges did they face, and how did they handle them? What, if anything, would they have done differently? How would their experiences be helpful to your present work?

Another important point to keep in mind is that many students are apprehensive about a design project because the thought of doing engineering design with their limited technical training can be overwhelming. These feelings can be especially strong for entry-level engineering students, and you should take them into account when discussing your expectations. For instance, you can temper your expectations about the technical sophistication of your designs based on the amount of science and engineering to which they have been exposed. An important part of any design course (especially entry level) is to learn about the iterative design *process* and the *skills* required to go back and do it better.

Remember also that a "negative" outcome, such as a design that is not technically or economically feasible, does not always represent failure. In the day-to-day world of engineering, a decision not to pursue a project further can be just as important as the decision to move ahead with production irrespective of other consequences. On the other hand, although you should not get hung up expecting technical sophistication, there are a number of things you should expect from yourself and your team members. Your instructor most certainly will. You should expect:

- Everyone to work hard and to work together productively
- Everyone to participate in the design process
- Everyone to do quality work (e.g., be thorough, write well, make accurate technical presentations)

- Everyone to meet deadlines
- To apply whatever technical/engineering skills you possess
- To learn some things about the technical aspects of your design topic.

Even these expectations can be defined in more specific terms. Plan on discussing them with your team members. For instance, clarify what it means to work hard. The guidelines for establishing a working agreement in Section 3.1.3 can help you define these and other expectations more completely.

You and the members of your team also may want to write a brief statement of objectives or mission statement. This is a brief statement describing the goals your team hopes to achieve. It should be less than 100 words, should reflect the consensus of the team, and should be written so that someone outside the team can understand it. Here is an example of what one might look like:

> *The purpose of this team is to support one another's efforts to learn about the process of engineering design by working collaboratively and to the best of our abilities. Our goals are to learn by creating an innovative design that meets functional requirements and realistically considers constraints and limitations. We will complete our work in a timely manner and strive to obtain an "A" in the course.*

One way to create your team's statement is by first having each team member write the statement individually. Next, discuss as a team what each of you has written and reach a consensus on a team statement. You also may want to refer to course objectives provided by your instructor as a starting point.

3.1.2 Coordinating Schedules and Planning Meetings

Even if your team has developed a clear and cohesive idea of what you want to accomplish, you will have a hard time being successful if you do not make sure you have time to do the work. It is more than likely that you will see your team members on a regular basis as part of your design class, but you probably will also need to make plans for meeting outside of class. Therefore, discuss when and where you will meet and expect to modify meeting dates and times depending on where you are in the design process.

Take some time to learn about the other obligations and commitments your fellow team members may have. Like you, your team members are probably juggling numerous things at once, from course work to jobs, from extracurricular activities to family responsibilities, not to mention time for socializing. You and your team members cannot allow other activities to become excuses for neglecting your design project. At the same time, you will be more likely to create a collaborative team atmosphere if you at least demonstrate sensitivity to the many demands upon one another's time. By having such discussions you also may find that you are able to help one another in other ways (e.g., ride sharing, study groups, etc.). Listed below are some additional points to keep in mind when coordinating schedules and planning meetings:

- Select times and places that work reasonably well for everyone. People are less likely to be focused if they are meeting at a time or in a place that is inconvenient. On the one hand, people should be flexible and try to accommodate one another; on the other hand, people need to speak up if a day or time is not convenient. Morning people and night people often compromise by meeting in late afternoon.

- Try to create additional incentives for attending meetings. We know of one design group that decided the best time for all of them to meet was early in the morning before class. To make meeting at this time more appealing, they established a rule requiring that they would take turns bringing breakfast.
- Avoid meeting at times or in places where people are likely to be distracted. Friday night meetings at the fraternity house are probably not a good idea. We have tried it ourselves, and, although we had a good time, we failed to get much done.
- Establish specific starting and ending times for your meetings and stick to them. It may take a while before you and your team members have a clear idea of how much meeting time is necessary. It can help to ask your professor for a general idea, and, if possible, talk with others who have already completed the course.
- Make sure you have clear reasons and objectives for meeting. It also helps to have a good idea what you want to accomplish by the end of the meeting. We suggest a posted agenda and action items.
- Do not schedule meetings unless they are really necessary. Meetings for the sake of meeting can eat into productive subgroup work.
- Establish ground rules for your meetings as part of your working agreement.

It also helps to have to have a general idea of your responsibilities as a team member in relation to effective meetings. The guidelines in Table 3.3 describe these responsibilities in general terms. You will have to work with your team members to execute them in ways that are specific to your team and its tasks.

TABLE 3.3 Team Member Responsibilities for Ensuring Effective Meetings

Before meetings	During meetings	End of meetings
• Identify agenda items and issues to discuss at meetings, and prepare a brief agenda in advance. • Complete any work or assignments from the last meeting before attending the next meeting. • Plan to arrive on time and stay until the end. • Come prepared to contribute and participate.	• Support and encourage one another. • Share ideas and information and participate fully. • Clarify misunderstandings and avoid unnecessary conflicts. • Listen actively to one another and ask questions when you are not clear or are confused. • Adhere to the agenda/objectives.	• Summarize meeting outcomes and next steps (action items). • Accept follow-up assignments and work. • Reflect upon how well the team is working together and suggest ways to improve. • Close with a clear idea of when and where to meet next.

3.1.3 Establishing Working Agreements

Whether you do so deliberately or not, your team's first few meetings are when people begin to form impressions of one another and establish expectations about how they will work together. These expectations, or *norms* as they are sometimes called, are standardized patterns of behavior that define what people expect from one another and what is considered acceptable and unacceptable behavior. Therefore, in addition to attending to *what* your team works on (e.g., the design problem you solve) you need to give serious thought to *how* your team works together. The best time to do this is when your team first begins to meet. (You will learn in Section 3.3.1 that this period is referred to as the forming stage of your team's existence.)

There are two ways norms can form in a group. They can be *implicit*; that is, they establish themselves based on what people do over time. The problem with letting norms form *implicitly* is that poor patterns of behavior may emerge. For example, have you ever been part of a group where people consistently arrive late or tend to hold annoying side conversations during group discussions?

The way to prevent unproductive behaviors from becoming the norm is to establish *explicit* productive norms. In these instances norms are established because people take the time to discuss what it is they expect from one another in terms of behavior and performance.

A working agreement is a list of common ground rules that define how a team will work together. In other words, a team's working agreement is essentially a list of its desired norms and expectations from one another. Team members may not always live up to this agreement, but they have a much better chance of doing so when everyone has the same shared understanding of what is expected.

Your Team's Norms At this point, you may be wondering what kinds of things to include in a working agreement. A good place to start is by reflecting on the norms that currently exist within your team. Examples of productive and counterproductive norms are listed in Table 3.4. Review these lists and place a check next to those that apply to your team.

Preparing Your Team's Working Agreement Now that you have spent some time reflecting on the current norms at work in your team, it is time to prepare your own working agreement. Begin by reviewing our checklist of team norms with your team members. Together, determine which of the norms on the list above you want to establish or enforce and which you want to eliminate from your team. Also, consider any other norms not on the list that you would like to add for your team.

To create your working agreement, identify principles that best represent the norms you want to reinforce and those you want to eliminate. In some cases, you may want to copy norms verbatim from the list. In other instances, you might create overarching statements that encompass several norms. You don't want to establish too few norms so as to create problems down the road or too many so that big principles are overlooked. Ten to 20 principles is a good rule of thumb but your team may have more or less depending on your needs.

TABLE 3.4 Team Norms

Productive norms	Counterproductive norms
▪ We have specific and clear goals, milestones, and deadlines. ▪ We try to learn from one another. ▪ Everyone understands what is expected of each team member. ▪ People carry out their assignments cheerfully. ▪ Leadership is shared. ▪ We usually consider several alternatives before we make major decisions. ▪ We treat each other with respect, especially when we disagree. ▪ Different opinions are valued. ▪ When problems arise we deal with people directly and not behind their backs. ▪ Meetings start and end on time. ▪ People are good listeners—we try to understand before being understood. ▪ We focus on problems, not people. ▪ If we don't like an idea, we try to offer an alternative. ▪ We try to solicit input from one another. ▪ Everyone has an obligation to participate. ▪ We usually have an agenda/plan for our meetings. ▪ We keep good records of what we did and when we did it. ▪ We try to have fun. ▪ We always try to reach consensus on important decisions.	▪ It is acceptable to miss deadlines. ▪ People pretty much do what they want, when they want. ▪ The same people consistently do most of the work. ▪ One or two people usually tell the others what to do. ▪ We usually accept the very first solution that sounds reasonable. ▪ It is best to avoid conflicts. ▪ We vote on *all* decisions, even small ones. ▪ If some do not speak up for themselves their views do not get considered. ▪ Sometimes it is okay to get personal when disagreeing or critiquing another's ideas. ▪ Side conversations are okay at meetings. ▪ Our meetings are usually much longer/shorter than planned. ▪ During meetings/discussions some people go off on tangents or talk at length about unrelated topics. ▪ Our records are such that no one could replicate our results. ▪ No one seems to be having fun . . . just drudgery. ▪ We rarely achieve consensus and decisions are rarely made.

All team members should feel comfortable with the working agreement. By all means, your working agreement should provide guidance relating to the following kinds of issues:

- Communication
- Participation
- Decision making and problem solving
- Managing disagreements
- Responsibilities and expectations

Working Agreement—Team A	Working Agreement—Team B
1. Show up to class meetings. Be on time. Let others know about conflicts. 2. Respect others. Don't take criticism personally. Criticize ideas, not people. 3. Share the work load evenly. Let others participate. 4. Communicate clearly and constructively. 5. Stay on topic and avoid distractions. 6. Work out disagreements. Give everyone a fair say. There is no rank in the group. 7. Be responsible for getting your work done on time. No surprises. 8. Come prepared to work. 9. Make sure everyone gets their journal in on time, each time.	Group Organization ▪ Two group meetings to be scheduled before the end of class every Monday ▪ Check email regularly and stay in contact with group members ▪ Minimize absences and inform group members prior to absence Meeting Format ▪ Focus on current task and work to accomplish it before moving on to other business ▪ Work efficiently during group meetings and minimize tangent conversations ▪ Major decisions always involve three or four group members present ▪ Keep group members informed as to all important issues and concerns – no surprises ▪ Any divisive issues are to be discussed thoroughly before any voting is done ▪ Document all group discussions and collective ideas General Guidelines ▪ Address personal or work problems as they arise ▪ Always turn in journals and remind other team members to do so ▪ Divide work fairly and discuss any problems ▪ Arrive at class meetings on time and be prepared to work ▪ Be open to all new ideas ▪ Keep good documentation of your work and distribute it to others ▪ This team agreement is open to modification and will be reviewed regularly

FIGURE 3.1 Examples of working agreements from actual student design teams.

- Conduct during meetings
- Conduct between meetings.

Examples of working agreements for student teams are provided in Figure 3.1. Which of these agreements do you think does a better job of addressing the issues in the checklist?

3.1.4 Adhering to Your Working Agreement

Once you have created a working agreement, you need to make sure it is put to use. You can do several things to make sure the agreement is adhered to.

1. *Establish structure for your meetings.* Consider assigning team members distinct "process" roles during meetings and rotate these roles among members from time to time. One person should be a timekeeper and keep track of how much time is spent on various topics. Another person should be a note taker. Note takers do not need to keep verbatim records. They should focus on recording the main topics discussed, key decisions reached, and action steps to be taken. Another important role is that of facilitator. At least one team member at each meeting should assume primary responsibility for monitoring your team's behavior in relation to its working agreement.
2. *Have all team members serve as secondary facilitators.* In addition to having a primary facilitator, all team members should assume responsibility for ensuring that the working agreement is adhered to. Make sure every member of your team has a copy of the agreement and is familiar with its contents.
3. *Assess your team relative to your working agreement.* It is a good idea to periodically review your team's performance in relation to your agreement. During these reviews the team should collectively explain what they are doing well and explore ways they can be even more effective.
4. *Build your own skills.* To a large extent your team's ability to work well together depends on each member's team skills. Therefore, identify specific skill areas you want to improve (e.g., active listening, conflict management, etc.). Review the relevant material in this text and establish a personal improvement plan. Ask your fellow team members to help you by monitoring your behavior in relation to your own goals.

3.2 COMMUNICATION

We often think of communication skills as being about the ability to convey ideas to others. Indeed, this is an important part of communication. In this section, however, we want to stress another aspect of communication first—the ability to understand the thoughts and ideas that other people are trying to convey. Specifically, we want to introduce you to techniques for active listening and probing. These techniques are useful for two important tasks. The first task is obtaining information and facts in order to more clearly understand your design problem's functional requirements and constraints and limitations. The second task is building relationships and rapport with other members of your design team.

Written communication is also essential for design engineers. Therefore, this section also provides a technique for documenting your work and reviews some important concepts of basic composition.

3.2.1 Active Listening and Probing Skills

As we indicated previously, engineering is much more than just applying scientific principles. Scientific principles are critical, but engineers must solve real problems for real people in real time. This means communicating with others to understand their needs and constraints and, at the same time, conveying your point of view in a way that is clear and meaningful. Your ability to listen well is one of the keys to clearly defining a problem and working successfully with other members of your design team.

Have you ever participated in a conversation in which participants were so busy expressing their own thoughts that they were not really responding to one another? After a while, perhaps one or two people dominated the conversation so much that others stopped talking altogether, feeling as though there were no point in trying to contribute "since no one was listening anyway." As a result, some good ideas were inevitably excluded from consideration. At the very least, the group probably missed the opportunity to build on one another's ideas and to capitalize on the synergies of their collective efforts.

In some cases, people may "mentally" drop out, not just from the conversation but from the group altogether. Afterward, perhaps you were surprised to learn that you had one notion of what was agreed during a conversation but others had very different recollections. At that point you may have wondered "How could we have all been in the same place at the same time and have such different ideas about what was said?" The reality is that there are many examples of technically competent people who make bad decisions because they fail to listen.

Listening may seem like an easy thing to do. In reality, though, it takes concerted effort, which is why we call it *active* listening. Active listening means you are conveying to the other participants that you are attempting to understand their point of view and that you are using your role as a listener to move a conversation forward. When you are listening actively, you help establish a participative environment for your team. Team members believe their contributions to the team are valued, mutual trust improves, and people build on one another's suggestions to create new possibilities. Misunderstandings are minimized, and, as a result, unnecessary conflicts are avoided.

Being an active listener (Table 3.5) requires that you do the following things:

- Pay attention.
- Deliberately focus on understanding, not responding.
- Attempt to move a conversation forward by showing your understanding.
- Help others expand on their ideas.
- Show support (not necessarily agreement).
- Probe for more information.

The way you respond to others indicates not only whether you heard them but also whether you understand their viewpoints. People are more likely to stay focused on the problem to be solved when they think you have not only heard their words but

TABLE 3.5 Some General Guides for Effective Listening

What	How
Let the other person in.	Give full attention, block distractions, demonstrate sincerity.
Listen to understand.	Respond to feelings and attitudes, hear the person out, watch nonverbal signs.
Listen for ideas.	Listen for facts that support key ideas, judge content not delivery.
Keep emotions under control.	Respect other views, set aside prejudices, avoid jumping to conclusions.
Participate mentally.	Ask questions, restate what you hear in your own words.
Respond.	Keep good eye contact, provide feedback.

have also demonstrated to them that you really understand what they mean. In fact, sometimes the way you respond will actually help others clarify what they are trying to say.

There are five broad categories of responses you as a listener can provide. They are described by experts as evaluative, reflective, supportive, probing, and summarizing. You may be called on to listen in one way or in all five ways at once in a meeting.

1. *Evaluative responses:* These send a message to others that you are judging what they are saying in terms of correctness or relative goodness. Although evaluative responses are not necessarily bad, they should be presented carefully. Because they are judgmental in nature, evaluative responses can alienate others and lead to a breakdown in communication. For example, "That's a dumb idea!" or "That idea will never work!"
2. *Reflective responses:* These responses reflect back to people how you interpret what they are saying. Reflective responses share with the speaker how the message was understood in terms of content and/or feelings (e.g., "What I hear you saying is . . . ").
3. *Supportive responses:* These indicate your intent to reassure other participants or to reduce their anxiety. The intention here is to help the other person relax. It is important to note that, although you are showing concern, you are not necessarily agreeing.
4. *Probing responses:* These responses are designed to seek additional information, data, and details or to provoke further thought. One of the best ways to probe is by using open-ended questions that begin with words like "Why?" or "How?" These probing questions encourage people to elaborate on what they are saying. In contrast, closed questions can be answered flatly with "yes" or "no" responses. Most closed questions can be rephrased as open-ended questions for greater detail or explanation.
5. *Summarizing responses:* From time to time it is a good idea to organize and restate what has been said thus far in a conversation. Doing this can be very helpful

The Design Situation: You and the members of your design team have been asked to develop a new grocery cart that will separate breakables from boxed items. You have been trying to clarify this problem for several hours but seem to be getting nowhere. Finally, a member of your design team declares angrily, "This project is a waste of time. We just keep going around in circles and we can't get anything done."

Evaluative Response—"Well maybe if you stopped complaining and tried to participate we *could* accomplish something."

Reflective Response—"You seem concerned that we won't be able to clearly define this problem in a way that leads to a viable design."

Supportive Response—"I can appreciate your concern. After all, a lot is riding on the success of the design we come up with."

Probing Response—"What do you think might be some better ways for us to approach discussing this problem?"

Summarizing Response—"This is a good point to review what we have considered up until now. It seems to me we have been discussing three different issues. Is that correct?"

FIGURE 3.2 A student design problem scenario.

when people seem to be getting stuck or drifting off the topic (digressing). Attempts to summarize usually should be followed with a question to make sure you "got it right."

Figure 3.2 describes a likely scenario you may encounter while working on your design problem. Following the scenario are examples of different types of listening responses. Which response(s) do you think would be the most helpful to this team at their present stage of progress?

You should recognize that the evaluative response in Figure 3.2 is likely to inflame the team member's frustration. You may actually believe the other person *is* complaining and not participating. But remember, your first goal as an active listener is to convey an understanding of the other person's viewpoint. By doing so you stand a better chance of eventually turning this person's attitude around. The reflective and supportive responses convey that understanding and, when combined with the probing or summarizing responses, help direct everyone back to the work to be done.

At this point you may be thinking "What about the fact this team member's statement and lack of participation is annoying me? Don't I have a right to let him or her know how I feel?" The answer is yes, but remember that you want to do so in a way that solves the problem. Consider a response like the following,

I am getting frustrated too but I am concerned that if we do not stay focused things will not get any better. That is why I could really use your help and attention. What do you think might be some better ways to approach our work right now?

TABLE 3.6 Assessing the Listening Effectiveness of Your Team

If you have already begun to meet with your design team, perhaps you have a sense of whether people listen actively to one another. Can you think of specific things that certain team members did to convey to one another that they were listening? For each active listening technique, try to provide an example from your team. If you can't think of something someone actually did, try to think of an instance when the technique might have helped:

Asked open-ended questions (e.g., What, How, Why?).

Restated or paraphrased what someone else said.

Solicited input from one another and encouraged alternative points of view.

Summarized the group's discussion to ensure mutual understanding.

Used nonverbal gestures to help convey that they were listening (e.g., nodding, maintaining eye contact).

Used reflective responses to convey an understanding of someone's feelings.

Showed support for other team members even though they may have disagreed.

This response conveys your understanding but directs the other person toward what you would like him or her to do differently.

Finally, take a look at Table 3.6. Use it as a guide to reflect on the listening skills you and your design team members are currently using.

3.2.2 The Laboratory Record Book

The very first bit of writing by an engineering professional, as well as the freshman design student, is usually done in the laboratory record book (LRB), also known as the laboratory notebook or project record book. The engineering professional uses a bound notebook with preprinted, consecutive page numbers to prevent tampering, adding, or deleting pages. The Battele Memorial Institute, a private science research corporation in Ohio, as well as many well-known major corporations require a well-kept LRB for all their associates. Properly signed and dated LRBs can serve as legal evidence in a court of law if a dispute arises, and they are used to support patent applications. The engineering design student can view the LRB in a similar way: to provide evidence and documentation of the learning and the design process. Instead of a bound notebook, however, the design students more often will keep a running three-ring binder so they can add materials from the professor and other design team members. Figure 3.3 lists the contents of an LRB prepared by students working on the autonomous robot design project.

Section	Contents
1	Design Schedule and Team Working Agreement
	Original Schedule
	Revisions to Schedule
	Working Agreement
2	Budget
	Part Order Form
	Standard Parts List w/ costs
	Interim Budgets
	Final Budget
3	Handouts
	Course Syllabus
	Project Statement and Specifications
	Drawing of Robot Course
4	Code
	Functions
	Final Code for Competition
	Information from Hands-On Labs
	Sensors w/ Code
5	Sketches
	Chassis
	Drive Train
	Pickup Mechanism
6	Final Drawing
	Chassis w/ Drive Train
	Pickup Mechanism
7	Calculations
	Motor Calculations – Power
	Course Calculations – Distance / Timing
8	Sensors and Controller
	Electrical Wiring
9	Final Report
	Outline
	First Draft
10	Oral Report
	Outline
	Draft of Visual Aids
11	Team Notes on Problems, Solutions, Decisions
	Problems and Solutions
	Key Decisions
12	Behavioral Checklist
	Phase I
	Phase II
	Phase III
	Phase IV

FIGURE 3.3 Table of contents for a student design team laboratory record book.

Central to the LRB is "the problem statement" introduced in Section 2.1. Placed prominently in the student's lab notebook, it keeps the engineering student focused on the need for a design solution. The rest of the LRB consists of activities, fully referenced and dated, at each step of the design process. The LRB should be able to be used by a subsequent engineering design student to replicate the design and to avoid mistakes, dead ends, and failures. Some engineering design professors grade LRBs at the end of the quarter or semester in terms of neatness, completeness, structure, and usefulness for someone else to duplicate the efforts of that team member. Regardless of whether the LRB is graded, the contents of a well-crafted notebook feed directly into the writing of a design proposal, discussed in the next chapter. For now, let us concentrate on basic composition.

3.2.3 Composition Skills

At a recent workshop on engineering writing and professional communications centers, the dean of engineering at a state university astonished some of those attending when he challenged the general concept of "technical writing," as a form of writing that is dreary and drab. "Technical writing," he said, "has two basic rules—be concise, and be technical... Write in the passive voice. Write in the third person. Do not use color. And for heaven's sake, do not add style." The result is dry, lifeless communication. Professors who write such prose are depicted as Mr. Magoo, and students with such technical writing skills are depicted as Dilbert—both equipped with near-sightedness and pocket protectors. However, the dean asserted: "We value good writing, and it comes in many grades of quality. But the writing of engineers need not be poor, need not be sterile, and need not be uninspired." He added: "I do not expect to find the instruction manual for my VCR to read like Wordsworth, and certainly not like Faulkner, but I do expect good writing—technical or otherwise."

Why does this dean expect good writing, even from his engineering students? He explains: "The leaders of the future, as in the past, must be articulate, thoughtful, liberally educated, and well skilled in the art of communication. Engineers must be so—and even more so than most. For it is engineers who will be responsible for educating our society on technology for the future, who will warn us of the dangers of technological wanderings, and who will create the knowledge and the design of products or services which will be the foundation for our economic strength in future generations."

Composition skills fall into three broad categories, ranging from surface errors to structural problems (Table 3.7). It may be a useless task to figure out which of the three areas is least or most important, because sloppiness in any one category can be seen as sloppiness of design or ideas. Therefore, each should be considered equally important:

1. Grammar, spelling, and punctuation
2. Structure, organization, and development
3. Style, syntax, and diction.

Each of these skill sets can be mastered first by understanding the terms.

TABLE 3.7 Common Composition Errors

The seven deadlies of composition
1. **Spelling errors** (especially names of people)
2. **Incomplete sentences/sentence fragments** (*Mistake*—"Like this." *Correction*—Revise the sentence to incorporate the fragment. "A complete sentence, like this, is allowed."
3. **Runon sentences** (*Mistake*—fusing two sentences together without punctuation. *Correction*—Separate the sentences.)
4. **Comma splices** (*Mistake*—fusing two sentences together with only a comma. *Correction*—Separate the sentences.)
5. **Subject—verb disagreement** (*Mistake*—"Everybody were there." *Correction*—"Everybody is there.")
6. **Noun—pronoun disagreement** (*Mistake*—"Everyone did *their* own compositions. *Correction*—Change the subject to a plural, "Students did their own compositions."
7. **Inclusive language violations** (Mistake—"Everyone did *his* own composition." *Correction*—Make the subject plural, "They all did their own compositions."

Grammar, Spelling, and Punctuation Grammar is best described as the science of composition. Grammar rules, such as noun/pronoun agreement and subject/verb agreement, are actually determined and altered by usage, not by grammarians or rhetoricians. Nevertheless, an educated professional is expected to learn, know, and practice them consistently. They are taught in elementary school but practice drill beyond that initial training has diminished. However, a new field of professional composition is emerging on major college campuses, and most universities now offer a course in intermediate writing or advanced composition. Consider taking such a course. The skills you pick up here will carry you well into a professional career and will become critical in management positions. Community and junior colleges are also beginning to specialize in remedial composition skills.

Spelling skills are deemphasized somewhat with the advent of electronic spellers and spellcheckers in virtually all word processing programs. However, none of them can detect wrong words or letter transpositions like from/form and no/on. Also watch for confusing words. Examples include accept/except, a lot/allot, affect/effect, cite/site, lead/led, loose/lose, medium/media, than/then, and principal/principle. We still need to read through our reports to catch errors. Reading aloud is a good technique for proofreading. Of course, we never misspell the name of our professor or client.

Punctuation has become somewhat automated by very imperfect grammar checkers. As a rule of thumb, do not use commas unless you have to, such as in a compound sentence. Generally avoid the semicolon and the exclamation point. Understatement is far more effective than overstatement, parentheses are more personal than semicolons, and the dash (not hyphen) is more modern than the semicolon. Learning just these few marks once is usually enough to be able to use them correctly.

Structure, Organization, and Development: Introduction, Body and Conclusion The structure of any document should be clear. The introduction should set the stage for your *main* point. Yes, every paper and report makes a point . . . otherwise it is pointless. That main point or thesis statement appears at the end of the first paragraph in a short report, after it has been introduced. You lead up to it. Rarely do you ever start with it. That's like hitting the reader over the head.

Each body paragraph should relate back directly to your main point. Call them subpoints if you wish, but they usually appear in the first sentence (often called the topic sentence) of each body paragraph. The rest of the paragraph *develops* the topic sentence, which, in turn, develops or proves the main point. The most common way to develop a body paragraph is to use examples, although you can also use evidence, proof, logic, persuasion, illustration, facts, and anecdotes to develop an idea.

The conclusion generally starts with a reiteration of the thesis or main point and concludes your discussion, finishing it off nicely. Your best strategy in the conclusion is to answer this question: So what? So you propose to design a workstation in your dorm room, so what? So you want to help a client decide if she should switch from oil to gas heat, so what? Show the implications or ramifications of such a notion. Then, end your report, don't just stop. Some writers like to use a "tie-back," echoing a thought from the introduction at the end of the conclusion.

Style, Syntax, and Diction If grammar, spelling, and punctuation are micro concerns, and if structure, organization, and development are macro concerns, then style, syntax, and diction are middle-range concerns for they deal with words and sentences.

Style is your imprint. You have a slightly unique way of presenting your ideas. This is particularly evident in public speaking, where you use inflections, gestures, volume, rate, and pronunciation to express yourself. But it is also evident in composition. An expert can tell which of two sentences, for example, belongs to Hemingway or to Faulkner.

On another level, style is a matter of common acceptance and expectation, and every profession seems to have its own way of presenting information. What style of documentation is preferred? Do we use in-text citations or footnotes?

Syntax is an entirely different matter. Syntax is basically sentence structure, and the basic sentence structure is S-V-O—subject-verb-object. Occasionally you may want to start a sentence with a phrase or a clause, and that is fine. But most sentences in a good, clear report are S-V-O instead of complex or compound-complex. Sentence length is also a concern. Some word processing systems indicate the grade level you are writing at, automatically as you compose. If the sentences are too long, it may take a Ph.D. to understand the writing, but if sentences are too short, a fifth-grader could read them, and you may be writing down to your readers. Most style indicators, like the Gunning fog index and the Rudolf Flesch formulas, suggest sentences that average about 20 words. Like the previous sentence.

The same style systems suggest a mix of words averaging about two syllables over the course of the manuscript. Diction is another word for word choice. In general, you want to choose simpler, shorter words over the longer, more sophisticated ones, if your aim is clarity. If you aim to impress, go ahead and use the big words, but expect to lose all but the headiest of readers.

We conclude with four rules of thumb.

1. Develop your own unique style, but make it natural, not pretentious or fake. You develop style only through practice. Keep a journal.

2. Keep sentences short and clear. Reread, or have someone else read, your material to discover the contortions and mental gyrations your reader would sometimes go through to read your most difficult syntax.
3. In terms of diction, select the right word (which is usually shorter) and not the big word. Use a thesaurus as well as a dictionary so you can compose with plenty of alternative word choices.
4. If you possibly can, tell a story (Table 3.8).

TABLE 3.8 Summary of Writing Tips

Ten tips for writers as storytellers
1. Know your subject Research and experience as much as you can on your chosen or assigned topic, but do not write beyond your area of expertise. Readers can detect whether a writer has a certain "feel" for the subject or just an academic knowledge of it.
2. Show, don't just tell Try not to block the window, to stand between the reader and your story. Let the reader see the story itself as you become almost transparent. Use concrete nouns and action verbs. Eschew adjectives.
3. Make a point If your story does not make a point, it is pointless, isn't it? If your story makes two or three points, you may have two or three stories, unless you subordinate all but one. The reader should detect a main theme running through the story from beginning to end.
4. Write to a title One good way to stick to a single main theme is to start with a clever, unique, insightful title. Put it atop each page to keep your story from drifting and losing the reader. Focus on it relentlessly, especially in a first draft.
5. Think multiple drafts Remember this is communication, not self-expression. Introduce the problem or conflict swiftly, build the suspense or excitement, and bring the story to a satisfying close. Repeat over and over until the development is apparent or obvious to the reader.
6. Make it endlessly interesting If storytelling is a boring, tedious task to you, imagine what your reader is going to feel. Put some heart into your story. Make us feel what you have felt. Pump yourself up, get enthused, and write with joy. Humor, too, is contagious.
7. Getting unstuck If and when you suffer from dryness or "writer's cramp," put your pen on paper and move it, or both hands on the keyboard and type something, anything, until the juices flow again. Half of all writers just gush it all at first and put shape to the story later.
8. Try an outline The other half ponders first, plots the full story out mentally or in a written outline, and only then begins to write. The first half may chose to outline the initial mess they created before revising it. Compose in the night, revise in the morning, or vice-versa.
9. Use a dictionary, thesaurus Fall in love all over again, this time with words. Check out a key word's etymology, read your second draft aloud. Find just the right word with just the right sound. Make friends with your software's spellchecker.
10. Boil, don't let it rise Today's reader wants your story tight, bright, well-organized, and crisp. Pare your story down to essential details. Write two or three short, snappy pieces rather than one long dreary one. Cut it until it begins to bleed.

3.3 COLLABORATION

You probably realize by now that engineering design involves more than just applying scientific principles. It also includes dealing with real people in the real world, including clients as well as your design team members. More often than not, it is a mistake to assume that interpersonal issues will take care of themselves. There are many examples of design teams that failed in spite of their outstanding technical abilities. As the cartoon in Figure 3.4 implies, successful teams rarely just happen. It takes concerted effort to establish and maintain a collaborative environment in which all parties to the design process contribute fully. In this section, we introduce you to some basic concepts that will help get your design team on the road to becoming a collaborative unit.

One of the first things to realize is that your design team, like all groups, will change over time. These changes include periods of uncertainty about the task and how to relate to your team members, periods of tension and frustration with what to do and how to do it, and fluctuations in productivity and performance. These changes are to a large extent time dependent—that is, your team's development into a truly collaborative design team typically involves going through several stages.

The specific number of and labels for these stages vary depending on which theorist or management specialist you talk to. We refer to them as forming, challenging, accepting, and collaborating. It helps to know what these stages are so that you can recognize them as they occur and better ensure that your team continues to grow and develop into a peak performing and cohesive unit.

"And so you just threw everything together? ... Mathews, a posse is something you have to *organize*."

FIGURE 3.4 Do not leave team development to chance. (Courtesy: Gary Larson/Universal Press Syndicate.)

3.3.1 Overview of the Stages of Team Development

In his 1981 Pulitzer Prize-winning book, *The Soul of a New Machine*, Tracy Kidder described how a team of engineers at Data General Corporation built a state of the art 32-bit minicomputer they called Eagle. As the title of the book implies, Kidder's story is less about the machine itself and more about the design team. The team accomplished their task in an amazingly short time and overcame many technical, organizational, and interpersonal challenges. Like all teams, this design team, known as the Eclipse Group, went through several stages, beginning with the forming stage.

1. *Forming:* This initial stage (Table 3.9) describes how people on a team struggle to understand their own personal interactive style, limitations, and capabilities while becoming comfortable working with other team members. This stage is characterized by a degree of hesitation and uncertainty among team members as they strive to become comfortable with one another and form their initial impressions. Kidder vividly describes what this stage was like for members of the Eclipse Group:

   Going to work for the Eclipse Group could be a tough way to start out in your profession. You set out for your first real job with all the loneliness and fear that attend new beginnings, drive east from Purdue or Northwestern or Wisconsin, up from Missouri or west from MIT, and before you've learned to find your way to work without a road map, you're sitting in a tiny cubicle or, even worse, in an office like the one dubbed the Micropit, along with three other new recruits, your knees practically touching theirs; and though lacking all privacy and quiet, though it's a job you've never really done before, you are told you have almost no time at all in which to master a virtual encyclopedia of technical detail and to start producing crucial pieces of a crucial new machine. And you want to make a good impression. . . .

   Eventually team members become better acquainted with their individual tasks and with one another. As inhibitions subside, people begin exchanging information and start to think of themselves as part of a team rather than as isolated individuals. It may not take long for a team to reach this rudimentary level of interdependence, but their work is just beginning.

TABLE 3.9 The Forming Stage

Common behaviors	Common feelings
– Attempting to define tasks	– Anticipation
– Determining acceptable behavior	– Confusion
– Floundering about where to begin	– Anxiety
– Overdepending on a single leader or individual	– Impatience
– Tending to "dive" into solutions	– Fear
– Being too "polite"; not wanting to make waves	– Lack of confidence

TABLE 3.10 The Challenging Stage

Common behaviors	Common feelings
-Arguing among team members	-Resistant
-Showing defensiveness	-Rebellious
-Establishing unrealistic goals	-Defensive
-Questioning the credibility of others	-Frustrated
-Choosing sides	-Angry
-Passing blame	-Suspicious
-Being rude	-Overconfident

2. *Challenging:* The relative politeness of the forming stage eventually gives way to the challenging stage (Table 3.10). At this point team members genuinely begin to express their different views and opinions about both the team's tasks and how they will work together. Conflict (and even some divisiveness among team members) characterizes this stage. It is important to realize that, although this conflict may be difficult at times, it is a natural and necessary part of your team's development. The key is to keep the conflict constructive by focusing on issues and avoiding personal attacks.

   The Eclipse Group at Data General did its share of challenging, too. One of the most formidable of these challenges centered around whether the team's objectives were even realistic. The leader, Tom West, faced strong resistance from the system architect, an engineer named Steve Wallach. First, West had to convince Wallach that the project was worth pursuing. According to Wallach, "Eagle would be backward and messy. What a comedown working on it would be." West overcame his initial resistance by convincing Wallach that the project would be a true challenge to his abilities. But Wallach had other concerns. As a veteran engineer, he had seen too many good projects canceled in spite of their merits. And he was suspicious of general management's willingness to support their work. It was not until these suspicions were overcome that Wallach began to work in earnest and with enthusiasm as a team member.

3. *Accepting:* This is the point at which team members begin to reconcile differences and clarify perceptions about what they will do and how they will go about doing it (Table 3.11). The agreement on common standards of behavior and expectations for performance characterizes this stage. At this point, conflict is giving way to a growing sense of cohesion and camaraderie. Members are becoming more satisfied with their work and are growing in terms of their confidence that they can get the job done. They are also likely to feel an increased sense of commitment to the team and their common objectives. There is less tolerance for disagreement and an increased willingness to conform to others. Of course, this does not mean there are no more conflicts. Different opinions are always an inherent (and healthy) part of the design process. Once a team has reached the accepting stage, however, intermember conflicts are

TABLE 3.11 The Accepting Stage

Common behaviors	Common feeling
–Attempting to achieve harmony	–An increasing tendency to identify yourself with the team
–Expressing opinions more openly	–Cooperative
–Sharing information	–Enthusiasm
–Showing less resistance to team tasks	–Relief
–Learning the best ways to do things	–Still a little tentative
	–Growing confidence

less likely to be of the sort that threaten a team's stability and existence. The following excerpt from Kidder's book describes what the accepting stage was like for members of the Eclipse Group.

You're a microkid, like Jon Blau. You arrived that summer and now you've learned how to handle Trixie, the nickname given to the mainframe computer the team was using to develop the minicomputer. Your immediate boss, Chuck Holland, has given you a good overall picture of the microcode to be written, and he's broken down the total job into several smaller ones and has offered you your choice. You've decided you want to write the code for many of the arithmetic operations in Eagle's instruction set. You always liked math and feel that this will help you understand it in new, insightful ways. You've started working on your piece of the puzzle. You can see that it's a big job, but you know you can do it. Then one day you are sitting at your desk studying Booth's algorithm, a really nifty procedure for doing multiplication, when you are told there is a meeting. . . .

At the meeting you are told that Eagle must be designed and brought to life in six months. That won't be easy but the brass think you can do it. . . . When you leave the meeting you go right back to your desk, of course, and pick up Booth's algorithm. In a little while though, you feel you need a break. You look around for someone to share coffee with you. But everyone is working assiduously, peering into manuals and cathode-ray tubes. You go back to your reading. Then suddenly you feel it, like a little trickle of sweat down your back. 'I've gotta hurry,' you say to yourself, 'I've gotta get this reading done and write my code. This is just one little detail. There's hundreds of these. I better get this little piece of code done today.' Practically the next time you look up, it's midnight, but you've done what you set out to do. You leave the building thinking: This is life. Accomplishment. Challenges.

For members of the Eclipse Team, accepting involved agreeing to take on the hours of long, continuous work required to meet their deadline. Along the way, they also had to learn ways of doing things that had never been done before. Clearly, they felt a great deal of pressure, but they also displayed growing enthusiasm and reveled in their identity as a team.

4. *Collaborating:* Once a team has passed through the accepting stage, they can move to collaborating (Table 3.12). This is the point at which your team is

TABLE 3.12 The Collaborating Stage

Common behaviors	Common feelings
-Balancing contributions	-Satisfaction
-Focusing on goals and results	-Energetic
-Solving problems collectively	-Motivated
-Being able to reach consensus and closure	-Close affiliation
-Encouraging criticism and constructive conflict	-Confidence in each other's abilities
-Sharing accountability	-Sense of fulfillment
-Following through on commitments	-Trust
-Pushing for higher standards	-Confidence in one's own abilities

truly focused on performing tasks. Team members not only have a clear idea of what to do but also are engaged in working together to accomplish their objectives. They are effectively managing not only their tasks but also the way they work together. Teams that have mastered the collaborating stage are like a string quartet in which each musician is fully integrating the sounds from one instrument with the others to create a seemingly singular melody. For the Eclipse Group at Data General, the collaborating stage was characterized by high degree of self-management and mutual commitment as exemplified in the following passage:

Over in the Microteam, though never explicitly told to do so, Chuck Holland took on the job of organizing the microcoding job. Holland and Holberger mediated the deals between the Hardy Boys [hardware engineers] and the Microkids [software engineers], but in general the veterans let them work things out for themselves. The entire Eclipse Group, especially its managers, seemed to be operating on instinct.

The Eclipse Group achieved a great deal because of their effort and initiative. Team members shared a very implicit and internalized understanding of what they had to do. Not all teams, however, effectively reach this level of integrated effort. In fact, research on group dynamics suggests that most teams do not. Nonetheless, it is an ideal to strive toward. Keep the behaviors and attitudes listed in Table 3.12 in mind as benchmarks for you and your team to strive toward.

3.3.2 Working through Your Team's Forming Stage

Let's focus on the forming stage (Table 3.13) and see how your team is dealing with it. Your actions and behavior during this initial stage can have a very profound effect on your team's ability to mature into a fully collaborating unit like the Eclipse Group. There are specific things you and your team members can do to establish a good foundation for your future success. Keep the following behaviors in mind as you begin to meet with your design team.

TABLE 3.13 Tips for Getting through the Formation Stage of Team Development

Effective team behavior during the forming stage
• Use *active listening* to convey your understanding of and interest in what others are saying.
• Make a point of soliciting input from *all* members of your team. This will help ensure that all team members feel included from the very beginning.
• Avoid rushing to conclusions on how to *define* your design problem. Encourage your team to consider several perspectives before making any decisions.
• Acknowledge *other* contributions and let people know their input is valued.
• Make sure your team spends time establishing clear structure and a sense of direction. For instance, make sure everyone shares the *same* perceptions about your purpose and goals. Encourage discussions about priorities and ways to proceed.

CHAPTER REVIEW

In addition to clarifying the nature of your design problem, Phase 1 of the design process is also the point at which you begin structuring your work. This includes establishing your work plan and beginning to build your team into a collaborative unit. Do not underestimate these tasks. Without them, you will find your design experience to be, at best, frustrating, and, more likely, a failure (regardless of how well you defined your problem). On the other hand, spending initial time in planning and team building will help you realize the successful completion of your design.

In terms of project management, we have stressed the importance of clarifying expectations, creating working agreements, and coordinating your efforts. With regard to communication, we have described the role active listening plays in team building and in clarifying your design problem. We have also introduced you to the importance of keeping good records and to the value of good composition skills. Finally, we have described the concept of team development and provided you with some behavioral guidelines to help facilitate your team's transition into a high-performance unit.

To help you review this chapter, answer the review questions. In addition, review the behavioral skills checklist at the end of the chapter and use it to assess how well you are currently applying the skills and tools discussed in this chapter.

REVIEW QUESTIONS

1. How would you characterize the project management issues your team will face during this initial phase of the design process?

2. How do working agreements help a team complete an engineering design?

3. What is meant by stages of team development? What implications does this concept have for the success or failure of a design team?

4. What is active listening? How can you use this skill to help your design team accomplish its objectives?

5. Try to develop different kinds of listening responses and strategies for the situation described below.

> *You work for Safe-Place, Inc., an engineering firm that develops customized security systems for residences and small businesses. When you arrived at work you received a message from a potential customer. She called you because this is now the third time someone has broken into her business. She wants you to design an electronic safety system to protect the property.*

Use your listening skills to learn more about this problem. What are some listening responses that will help you to better understand this problem? What kinds of reflective and/or supportive responses will help you convey an understanding of her needs? How can a summarizing response help you? Can you think of some evaluative responses you might want to avoid?

6. What is a laboratory record book? How and why do engineers use it

DORM ROOM DESIGN PROBLEMS FOR CHAPTER 3: DEFINING THE PROBLEM (PROJECT AND PEOPLE SKILLS)

Assignment 3 (project management skills)

Prepare a schedule of the actions and activities that your team will have to carry out during the next week to complete the problem definition phase of your project. Include in this project plan which team members will be doing what tasks, which team members will work together on what projects, and which team members will provide backup for each task. Use word processor, spreadsheet, or project planning software for your schedule. Include the time estimated and then actual time required.

Assignment 4 (project management skills)

Your team has now performed several tasks and your team members probably are now finding that things aren't going as well as you would like. Take 30 to 60 minutes and write down a set of team guidelines for your team members so that you can successfully complete the task at hand. Review (without accusing each other) the things your team did well and things that did not go so well. Write the guidelines so that each team member understands what is required.

Assignment 5 (communication skills)

Interview at least six students who live in a dorm (three male and three female) about their needs for individual dorm rooms. Prepare the list of questions so that each of you who interviews uses the same set of questions. Allow for unanticipated information. Prepare a report that summarizes your team's findings. Include a written summary and a table of results.

Assignment 6 (communication skills)

Part A: Establish a laboratory record book for your project. Document all team activities along with key information you have gathered to date.
Part B: Based on the information that your team has gathered, rewrite your problem statement. Include all appropriate information in written and graphic form. Be sure to describe functional requirements as well as any constraints and limitations impacting your problem.

TABLE 3.14

Common behaviors	Common feelings
Attempting to define tasks	*Anticipation*
Description	Description
Actions	Actions
Recommendations	Recommendations
Determining acceptable behavior	*Confusion*
Description	Description
Actions	Actions
Recommendations	Recommendations
Floundering about where to begin	*Anxiety*
Description	Description
Actions	Actions
Recommendations	Recommendations
Overdependence on a single leader or individual	*Impatience*
Description	Description
Actions	Actions
Recommendations	Recommendations
Tendency to "dive" into solutions	*Fear*
Description	Description
Actions	Actions
Recommendations	Recommendations
Everyone is "polite"; don't want to make waves	
Description	
Actions	
Recommendations	

Assignment 7 (collaboration skills)

One of the best ways to learn from a new experience is to reflect back on what happened while you were going through it. Assuming your team has been working on the dorm room design problem, take some time to think about how it is doing in relation to the forming stage.[1] The common behaviors and feelings associated with this stage are listed in the Table 3.14 below.

BEHAVIORAL CHECKLIST FOR PHASE 1: DEFINING THE PROBLEM

Instructions: Review each behavioral statement in the first column and reflect on your experiences working with your current design team. Then use the scale below to rate your own effectiveness with regard to each behavior. Record your ratings in the second column. Use the third column to rate the effectiveness your team. You may want to discuss your ratings with the rest of your team. In addition, you can make copies of this form and ask your team members to rate you as well.[2]

[1] If you have not been assigned the dorm design problem, you still can use this exercise to reflect on how well your design team is working together on some other problem.

[2] Appendix A contains a development planning form you can use to establish improvement goals in relation to specific behaviors. Focus first on areas with the lowest ratings based on your self-assessment and/or any additional feedback you receive from team members or your instructor.

For each behavior and feeling try to:
a) Describe an example that is specific to you or your team.
b) Indicate what (if anything) you or your team did to deal with the behavior or feeling.
c) Provide recommendations for improving the situation even more, if possible.

Rating Scale: 1 = Never 2 = Rarely 3 = Sometimes 4 = Frequently 5 = Always
N = Does Not Apply

Decision making	**You**	**Your team**
1. Recognized constraints and limitations impacting the problem.		
2. Worked to define functional requirements in specific terms.		
3. Looked for biases and assumptions in order to clarify or redefine the problem.		
4. Did not rush to conclusions regarding solutions.		
5. Used objective facts and specific research to help define the problem.		
Project management	**You**	**Your team**
6. Helped the team clarify goals and objectives.		
7. Clearly conveyed performance expectations and standards.		
8. Helped the team to clarify individual roles and responsibilities.		
9. Demonstrated appropriate flexibility with regard to meeting times and places.		
10. Actively participated in creation of the team's working agreement.		
Communication	**You**	**Your team**
11. Consistently asked probing and open-ended questions to clarify issues and increase understanding.		
12. Restated and paraphrased what others had said.		
13. Gave full attention to others when they were speaking.		
14. The written problem statement was well organized with clear main point(s) and supporting information.		
15. Ensured that grammar, punctuation, and spelling of written documents were correct.		
Collaboration	**You**	**Your team**
16. Consistently solicited input from other team members.		
17. Acknowledged others' contributions and ideas respectfully.		
18. Recognized and responded to others' feelings and concerns.		
19. Was friendly toward others and sought to build rapport.		
20. Demonstrated patience with others.		

BIBLIOGRAPHY

DEBONO, E. *Lateral Thinking: Creativity Step by Step*. Harper-Perennial, New York, 1990.

DYM, C. L. *Engineering Design: A Synthesis of Views*. Cambridge University Press, New York, 1994.

FENTIMAN, A. *Team Design Projects for Beginning Engineering Students*. Technical Report #ETM-10-05-958, Gateway Engineering Education Coalition, Philadelphia, 1997.

FORSYTH, D. R. *Group Dynamics*, 2nd ed. Brooks/Cole, Pacific Grove, CA, 1990.

FRYE, E. *Engineering Problem Solving for Mathematics, Science, and Technology Education*. Dartmouth Project for Teaching Engineering Problem Solving, Hanover, NH, 1997.

KIDDER, T. *The Soul of a New Machine*. Avon Books, New York, 1981.

ROGERS, C. *Technical Writing*. Workshop on Engineering Writing and Professional Communications Centers, University of South Carolina, Columbia, S.C. 1996.

WEGGEL, R, ARMS, V., MAKUFKA, M., and MITCHELL, J. *Engineering Design for Freshmen*. Technical Report ETM-1058. Gateway Engineering Education Coalition, Philadelphia, 1998.

CHAPTER 4

DESIGN PHASE 2: FORMULATING SOLUTIONS—STEPS AND DECISION MAKING

Has your team successfully defined the engineering design problem at hand? If so, you are ready to move on to the next phase in the engineering design process—formulating solutions. However, as you and your team begin to formulate solutions, it is entirely possible that you or another team member may slowly realize or suddenly discover that the problem definition needs to be refined, better focused, or even discarded in favor of a new one.

Design teams must be both willing and able to revisit any of the steps in any phase of the engineering design process when new information or new perspectives may affect the work in progress. Engineering design usually has been, and likely will continue to be, an iterative process. With this caution in mind, we turn next to the steps in formulating solutions.

Notice that the emphasis here is on formulating solutions and not just a single solution. It should come naturally that, since we faced an open-ended problem in Phase 1, there will probably be multiple solutions. The key is to select those solutions that have the greatest potential. As before, this phase of the engineering design process requires carefully executed steps to formulate solutions. These steps, along with some decision-making tools for completing them, are discussed in this chapter. The first two shaded boxes in Table 4.1 list the topics covered. In Chapter 5, we turn your attention to how project management, communication, and collaboration skills apply to this second phase of the design process. For now, let's look more closely at the steps engineers take to formulate design solutions.

4.1 DEFINING DESIGN PARAMETERS

Design parameters are the features and characteristics of a design that enable it to meet its functional requirements. Defining design parameters can be a difficult task for student design teams. Industry usually relies on their customers to define the parameters for a new product or process. As we saw in Section 2.1.1, where an example of informational survey results for the assistive feeding device project were summarized, student design teams will find information collected from real users

TABLE 4.1 Overview of Design Phase 2: Formulating Solutions

Steps for formulating solutions	Skills and tools for formulating solutions			
	Decision making	Project management	Communication	Collaboration
1. Defining design parameters 2. Identifying alternatives 3. Evaluating and analyzing alternatives 4. Selecting a solution	• Innovation versus origination • Considering external factors • Brainstorming • Normal group and delphi techniques • Lateral thinking • Systematic decision grids • Force field analyses • Making estimates	• Preparing and using Gantt charts • PERT/CPM techniques • Establishing and maintaining records	• Sharing the data gathered • Writing proposals • Preparing bibliographies	• Ensuring open participation • Reaching consensus and building commitment • Managing conflict • Avoiding groupthink

quite beneficial in this part of the process. When students can work directly with real users or customers on their projects, they have a valuable resource in defining the parameters of an engineering design project.

In the case of the freshman robot design project, students have a fairly clear notion of the robot size parameters and the functional requirements as they work on their design because these parameters and requirements have been specified in the problem statement. However, they do not necessarily have a good concept of the parameters surrounding the power train or the computer code for the project. In addition, developing the computer code for the project can appear to be a daunting task unless the teams break down the actions the robot has to perform (functions) and where those actions take place (terrain).

Robot design teams that were successful found that they needed a starting function, a turning function, a measuring function (using the shaft encoders to determine distance), a time function to keep track of time, a tilt-sensing function (using a mercury switch), and other functions such as crossing a gap, picking up something, sorting something, and placing something.

Knowing the parameters and segmenting the computer code for verification and testing can save time and effort in the long run, resulting in a better project. Thus, teams should devise portions of the code for sections of the course. It is particularly important to be able to start and stop the code at the beginning and end of each section of the course so that the robot function can be tested for that particular section without waiting for the robot to perform successfully on a section that has already been proven.

4.1.1 Innovation Versus Origination

As you identify alternative design parameters, you may find yourself considering solutions that are unique and or ones that are only subtly different from what already exists. In today's world, we find both innovation (a modification to something that exists) and origination (something brand new). Both have their place. The tricky part is to determine what is appropriate for a particular project. At one extreme, keep in mind that design solutions do not have to be novel in order to be useful. You want to avoid getting trapped by the NIH (not invented here) syndrome. This is a tendency to reject solutions because they are not original or were thought up by someone else. Most successful designs are innovations—incremental changes to existing forms. At the same time, it never hurts to try and look at a problem from a completely new perspective. Avoid imposing solutions simply because they have worked before.

In many cases, design solutions contain both original ideas and innovations. In *All Corvettes Are Red*, much of the new model was actually brand new—it had components that were manufactured for the Corvette only and no other vehicle. In this case, the motor was an adaptation of the original small block Chevy pushrod engine, and the new model Corvette returned to the tradition of being powered by a motor with a pushrod/cam arrangement instead of dual overhead camshafts. However, the engine block was aluminum rather than cast iron, and there were several other components that were new to the Corvette.

In the case of the freshman robots, the Handy Board controller is available for purchase and can be used for a variety of applications. It is considered to be off-the-shelf. A variety of other controllers could have been used that would have many of the same functions of the Handy Board, such as the Basic Stamp. However, in this case, the students learned the C/C++ language and not BASIC, so the Handy Board is a good choice. First-year students would have neither the time nor the expertise (in most cases) to build their own controllers from scratch.

This is also true of the DC motors that are used to power the robots. The students had the experience of building a simple DC motor from parts in a hands-on laboratory session. However, this is a crude motor and would not be satisfactory for this project. The same thing might be said for many remaining components—microswitches, for example. The students could design and build a switch, but it does not make much sense to do this in terms of time and quality assurance when they can be obtained off-the-shelf.

The freshman students are responsible for building their own chassis. Although they are given an Erector set, they can choose other components and materials for the chassis. Many teams use the Erector set components because they are familiar. Some teams have originated new chassis components. They have used wood, steel tubing, and small-diameter polyvinyl chloride (PVC) pipe. All are low-cost items but may require special tools for the assembly.

When civil engineers are designing a new bridge, it is almost always done with standard beams and fastening technology. Seldom do they ask a steel mill or steel fabricator to make beams with different cross sections. The design itself will probably be original because almost no application in our infrastructure is exactly the same as any other application in terms of terrain and water depth. In Cincinnati, for example, there are many bridges that cross the Ohio River between Ohio and Kentucky. They were built over the past 140 years and are very different in appearance, in part because the application for each was different. Many are from standard steel components. Some are reinforced concrete. Some are designed for trains but most carry automobiles and trucks. Some are designed to carry a very high volume of traffic, and others carry much less.

Much of engineering design today consists of choosing the optimum assembly of standard components in order to arrive at a cost-effective solution. The personal computer (PC) industry, reviewed from an historical perspective, shows a variety of microprocessors that were used as the central processing unit during the early years. Intel, Motorola, Texas Instruments, and Zilog all produced chips that were used in thousands of machines. In most of these cases, early PCs had their own proprietary operating systems. Therefore, the computer applications often tended to be unique to the processor chip used, with one exception: BASIC was the programming language used on most of the machines, and, although not standard, it was at least very similar.

Today, we have progressed to the point that there are primarily two chip designs vying for the PC market: the Intel (and Intel clones like those from AMD or CYRIX), and the Motorola/IBM PowerPC chip set. Consequently, two operating systems now are sold with most PCs—Windows and the Apple Macintosh operating system. Yes, UNIX and Linux are available as an operating system from a variety of sources. They tend to be used by those who want higher power computers and who can develop their own programming needs. Most of the currently popular software applications are based on the use of either the Windows or Macintosh operating systems.

This is part of what makes designing a new PC that will challenge the present brands very difficult. Dell and Gateway have been successful, in part, not so much from innovation in computer technology as from marketing strategy. If you want one of their machines, you call the factory, specify the components and features, pay with a credit card, and wait (perhaps just overnight) for it to be shipped to your house. They have cut their costs by not having stores and salespeople. Are their machines better than one assembled by an engineering student from standard components? This is a question that you might want to discuss with your friends. An additional point to note is that successful new products need more than their technological base; they need appropriate marketing approaches.

4.1.2 Considering External Factors

In many instances, requirements imposed upon a design by codes, regulations, and/or the need for the design to function as a part of a broader social, economic, or mechanical system are an important determinant of design parameters. For instance, in the book *All Corvettes Are Red*, the design team is presented with a number of factors external to the design and assembly of a sports car. The government has dictated that vehicles must have air bags for safety; must burn fuel efficiently so as to lower pollution and conserve petroleum resources; must have headlights, taillights, and bumpers that will withstand a crash up to a certain speed; and must have a host of other things that would not be put on a car designed to be competitively raced. Nevertheless, a new vehicle must have all these features in order to be marketed in the United States. Almost all designs have such external factors that must be considered and taken into account.

Student design projects, especially those in national or regional competition, also need to consider external factors. For example, in the SunRayce competition (see sidebar), vehicles have to be equipped to travel on public highways safely with the standard equipment of lights, horn, and turn signals. Some projects, like the freshman robot project, face restrictions imposed only by the instructors through the problem scenario. With both the robot and the assistive feeding device projects, cost is an additional external factor to be considered. In the case of the freshman robot design, cost considerations are imposed somewhat artificially. For the feeding device, safety factors are important and cost factors arise externally from the information collected from real users (what the "market" would support).

The SunRayce

SunRayce is a biennial, intercollegiate competition to design, build, and race solar-powered cars in a 10-day, long-distance event. Begun in 1990, the race usually is held from June 20 through June 29 to coincide with the summer solstice (the time of year when the sun's intensity is greatest.) The course has varied over the years, but the distance covered is usually a little over 1,000 miles. The 1999 version of the race traveled from Washington, DC to Orlando, Florida.

The event's goals are to promote educational excellence, renewable energy, teamwork, and environmental awareness. SunRayce also champions the creative integration of technical and scientific expertise across a range of disciplines.

The winner is the solar car with the lowest cumulative elapsed time to complete the official course. A new average winning speed record of 43.29 mph was set in the last race by the Univeristy of Missouri-Rolla. Vehicles race during daylight hours, from 10 a.m. to 6:40 p.m. and gather at a common stopping point each night. Sunlight is the only external source of power allowed for propulsion. Only commercially available solar cells and high-performance batteries may be used. Seat belts, a horn, turn indicators, brake lights, and a rearview mirror must be functional for safety.

Forty student teams from universities and colleges all over North America competed in the 1999 race. Only 40 teams are allowed to race but more than that apply. To be among the 40, teams must first compete at regional qualifying events.

The student teams work long and hard to design their vehicles. The design challenges they consider include driver's safety, vehicle weight and durability, efficiency of all components, aerodynamics, rolling resistance, energy recovery systems, and energy storage systems. The design tools they use include computer-aided design, computational fluid dynamics, velocity prediction, and satellite tracking. In addition, advanced programs and systems like solar array testing are used to model each of the vehicles and to predict actual performance under race considerations.

The most powerful car does not always win, nor does the lightest, the one with the most battery energy, or the car with the biggest budget. Many times the winning strategy comes from a good understanding of energy management—how much energy the car is collecting, storing, and using to power itself.

In addition to worldwide recognition, awards include cash incentives for the top three team proposals; trophies for first-, second-, and third-place teams; and scholastic achievement awards for technical innovation, engineering excellence, artistic talents, teamwork, and good sportsmanship.

(*continued*)

Sponsors include General Motors, the U.S. Department of Energy and Electronic Data Systems (EDS). They provide coordination and financial underwriting of the entire SunRayce project. In addition, they provide technical, professional, and logistical assistance to the teams, including aerodynamic analysis and photorealistic image design.

Over 6,000 students from 175 engineering schools have participated in the event since its inception. Maybe next time you and your fellow students will be among them.

SunRayce Photos—Courtesy The Ohio State University.

SunRayce Photos (*Continued*)—Courtesy The Ohio State University.

4.2 IDENTIFYING ALTERNATIVES

When an engineering design team in industry is brought together, team members probably have already had a variety of experiences they can bring to the project at hand. As such, they have seen things that have worked in other cases and probably have ideas about to what works, what doesn't, and what works but would work better if it were modified. In these situations, identifying alternatives is probably a matter of weeding out the ones that are proven to work or that may work from those that do not work or do not quite meet the functional requirements of the new project as defined.

This scenario is not necessarily true of student design teams. In particular, a freshman design team may have students who have never done an open-ended problem or design before. They need to draw on limited experiences from high school or from the college courses that they have taken or are taking as freshmen. Today, many freshman engineering programs provide hands-on laboratory exercises that explore how things work. These experiences allow them to define alternatives for design projects. It is possible that the team members will not see the connections immediately, but eventually most do see how their experience and expertise can be put to use.

Design projects for juniors and seniors provide opportunities for the students to use the analytical tools that they have been developing in their major courses. Many juniors and seniors have participated in cooperative education or internship programs that provide experience in real-world situations, and the cumulative experiences of the team members can provide the background that often is missing from freshman design teams.

Regardless of your skill and experience level, whether you are brand new to engineering, a more advanced student, or even a practicing professional, if you are doing engineering design you will be identifying alternatives and choosing among them. In other words, you will be decision making and there are certain tools an techniques that you will want to apply. In terms of generating alternatives there are three techniques that are particularly helpful: brainstorming, lateral thinking, and the Delphi and nominal group techniques.

4.2.1 Brainstorming

What has come to be called brainstorming is a good way to flush out and capture ideas that will be extremely creative, often revolutionary, and even imaginatively wild. It is precisely at this point when you are identifying alternatives that engineering design can become exciting because of the creativity involved in generating new ideas for new (or even old) problems. For engineers, who may not be able to sing songs, compose music, write poetry, or paint with oils, their own creative, artistic expression may well be found in the products or systems they design to solve a problem or meet a need.

Brainstorming is a method or technique by which members of a group can generate as many ideas for solving the problem at hand as possible in a relatively short period of time. It is one of many different strategies for overcoming barriers to creativity. The method has been around since the late 1930s[1] and both the term and the process are commonly used; however, not everyone is familiar with the way a successful brainstorming session should be conducted. Brainstorming sessions can be very successful if a few guidelines are followed.

1. *Criticism is ruled out.* Any judgment or evaluation of an idea must be held off until later. One of the goals of a brainstorming session is to generate as many ideas as possible in the allotted time; thus taking time to critically evaluate an idea means there will be less time available for generating other ideas. No idea should be squelched in a brainstorming session, because some people may hesitate to make suggestions they think may be too silly or impractical. This leads us directly to the next guideline.
2. *Creative and imaginative thinking is encouraged.* Participants should be free-wheeling, and even wild ideas are welcomed. A currently popular phase of encouragement in this regard is to "think outside of the box." Often an apparently wild or silly idea or concept may lead to the very solution that has been long sought. Such a solution may evolve from an offhand comment or from a simple idea initially passed off as a joke.
3. *Quantity is the metric.* The success of a brainstorming session is best measured in the quantity of ideas generated and not the quality. The challenge is to generate as many ideas as possible in the agreed upon amount of time. If a large number of ideas or concepts can be generated, then at least some of them will likely be promising—and generating some promising alternatives is the whole point. It is important to remember that this brainstorming process may produce not

[1] Alex F. Osborne. *Applied Imagination*, Scribner's, New York, 1957.

well-formulated final design concepts but merely promising ideas that will need much further refinement and development.

4. *Combining and extending are good.* Interaction among those in the session is key to combining and extending ideas that are put forth. Often a suggestion by one person stimulates formation of ideas that may be related in the minds of the other participants. This extension of one idea or a combination of two or more leads to idea refinement and improvement. And those ideas that are improved upon may be more promising.

Did you notice earlier that brainstorming involved members of a group? Did you naturally assume that this group was just your design team? A brainstorming session may very well be limited to those on the design team, but some diversity may also prove very beneficial. A group should be limited to perhaps a dozen or so participants, but, in addition to individuals from the design team, some people may be included who have little knowledge of the subject. A fresh perspective from someone who is not involved with the problem can provide remarkably good insights. Often in industry the group might consist of engineers from various disciplines together with managers, salespeople, production personnel, maintenance and service people, and distributors. Some suggest that managers and supervisors not be included in a brainstorming session so that the flow of ideas is not restricted. Whatever the composition of the group, the members should be provided with some information about the problem at hand at least several days before the session is held in order to allow some incubation time for ideas.

Brainstorming works better if the problem is sharply defined. Phase 1 of the design process has produced such a problem definition for you. At the beginning of the session the group should select one from among themselves to act as the moderator and one to act as secretary or recorder. The moderator's duty is to manage the session while the secretary takes down the ideas presented. After briefly describing the problem, the moderator recognizes in turn each member who has something to contribute. Remember that the interaction among group members is key, and ideas should be extended or combined. The session should move along fairly briskly and end naturally when the flow of ideas has slowed to a nonproductive rate. Some may prefer to place a time limit on the session, allowing anywhere from a few minutes to an hour or more. The secretary should provide a list of the ideas collected to the participants. Sometimes flip charts or blackboards provide an opportunity for all the ideas to be visible as the session continues. The list might have as many as a hundred ideas collected in a simple 20-minute session. It is up to the design team to pare down these ideas to the most promising ones that will be selected for further development and refinement

4.2.2 Lateral Thinking

Another way to generate ideas is through lateral thinking. This is not so much a specific technique as it is a general approach to considering a problem and its potential solutions. Lateral thinking stands in contrast to vertical thinking. Therefore, in order to understand *lateral* thinking it helps to first know what *vertical* thinking means.

Vertical thinking refers to the way people typically approach problem solving. It implies moving down a solution path by evaluating information logically and

TABLE 4.2 A Summary of Differences Between Lateral Thinking and Vertical Thinking

Vertical thinking	Lateral thinking
▪ has selecting an idea as a goal	▪ has generating ideas as a goal
▪ focuses on whether something is right or wrong	▪ does not concern itself with right or wrong
▪ is sequential	▪ jumps around
▪ is meant to be analytical	▪ is meant to be provacative
▪ excludes irrelevant information	▪ welcomes irrelevant information
▪ tries to finalize	▪ tries to expand possibilities

objectively. The point is to move forward in sequential steps, with each step justified by logic and fact. Along the way we evaluate new ideas in relation to existing ones and in relation to existing patterns and experiences. At this point, you may be saying, vertical thinking sounds pretty good. If so, you are correct. Vertical thinking is good. We need it to develop ideas and to ensure that designs really work. In fact, much of what is discussed in this text reflects the significance of vertical thinking (e.g. there are specific steps in the design process; you need to establish clear requirements and criteria for your designs).

On the other hand, vertical thinking has its limitations and that's where lateral thinking comes in. We stressed in Section 2.1.2 that human beings are creatures of habit. We look for patterns and relationships in order to make sense out of our world. The limitation of this pattern seeking is that in doing so we often fail to realize that there are alternative ways to look at a problem that could yield a new or even revolutionary solution. The purpose of lateral thinking is to help you break out of vertical patterns by reorganizing information and putting it together in different ways that can lead to new and unique ideas. According to Edward deBono, the originator of this approach, there are four critical factors associated with lateral thinking: 1) recognizing dominant assumptions that polarize perception of a problem, 2) searching for different ways to look at things, 3) relaxing rigid control of thinking, and 4) using chance to encourage other ideas. Table 4.2 summarizes some differences between vertical and lateral thinking.

We can easily fall into vertical thinking patterns even when using an idea-generation technique like brainstorming. For instance, good brainstorming sessions typically involve building off other people's ideas. This can be helpful but sometimes the result is a list of ideas that are only marginally different from one another.

In his book *Lateral Thinking*, deBono describes several techniques you can incorporate into brainstorming sessions and other idea-generation efforts. We already discussed one of these, the importance of *challenging assumptions* in Section 2.12, but in this section we expand on that topic a bit more. Remember the nine-dot problem in Section 2.1.2. To solve it we had to look past self-imposed boundaries that otherwise would keep us within a square. Once we dropped the assumption that we had to stay within a square, the problem became solvable by extending our lines past the square. When you are formulating solutions, it may help to identify the assumptions you are making so that you can see how they are affecting the ideas you generate. Consider the following brainteaser of Figure 4.1. To solve it you have to look past your assumptions.

A Lateral Thinking Puzzle

The teacher gave Ben and Jerry a written test. Ben read the test, then folded his arms and answered none of the questions. Jerry carefully wrote out good answers to the questions. When the time was up, Ben handed in a blank sheet of paper and Jerry handed in his work. The teacher gave Ben an A and Jerry a C. Why?

Do you have a solution? Here are some clues. Each boy deserved the grade he was given. There was something unusual about the test. Jerry was not as diligent as he should have been.

What assumptions have you made about the test? About Ben and Jerry? About the teacher?

FIGURE 4.1 Problem that requires looking past dominant assumptions.

Another lateral thinking technique is called the *reversal method*. To use this approach, take a design problem as it is and turn it around. For instance, suppose you were asked to design a device to hold a book over a bathtub so that someone can read in the tub without getting the book wet. You could conduct an initial brainstorming session to identify possible solutions. After a while, your group would begin to run out of ideas. To revitalize your thinking you could try the reversal method. Doing so turns your problem into something like designing a device to hold a bathtub over a book or keeping the tub dry while getting the book wet. Viewing the problem in this unorthodox way might prompt new possible solutions such as designing a waterproof book or creating a floating waterproof container in which to place the book while in the tub.

A third technique is called *random stimulation*. This approach involves using irrelevant cues or unrelated information as stimuli to prompt free associations and new ideas. For instance, some groups create a list of random words and then try to generate ideas by relating words from the list to the problem. As an example, try using some of the words in Table 4.3 to generate solutions to the design problem mentioned in the previous paragraph (designing a device to hold a book over the tub).

In summary, lateral thinking can help you be more creative and generate novel solutions to design problems. Do not, however, expect all the ideas you develop through lateral thinking to work. You still need to use vertical thinking to carefully evaluate ideas and establish a rationale for your final solution(s).

TABLE 4.3 Example of Random Word List to Promote Lateral Thinking

Random word list	
1. Weed	5. Magnify
2. Rust	6. Foam
3. Poor	7. Vacuum
4. Hole	8. Diagonal

4.2.3 Delphi and Nominal Group Techniques

These two techniques are methods for generating alternative ideas and are essentially variations of brainstorming. The difference is that these techniques do not necessarily gather people together into a single location, but you do gather their thoughts and ideas. Both techniques help you generate ideas and may also help you select an alternative. Here is how the Delphi technique works:

1. Select a panel of experts from class and outside of class, especially those who have completed excellent design projects already.
2. Ask each expert to comment *anonymously* on the topic at hand. The experts may propose a solution to a specific problem or predict the outcome of a scenario or event.
3. Feed each expert's anonymous solution back to the entire group.
4. Ask if anyone wants to change or modify any of the solutions.
5. Repeat until you arrive at consensus and come up with a composite solution.

The Delphi technique is well adapted to data collection by e-mail, as long as you can protect the anonymity of each expert. It is particularly useful when you want to ensure that strong personalities do not dominate a discussion and/or you need to avoid personality clashes. It also helps when you want input from people who are unable to attend a meeting. On the negative side, this process can take a long time because you have to repeatedly gather and disseminate information (as opposed to having people speak directly with one another). You also may lose out on some of the give and take learning that occurs when people talk face to face.

The example in Figure 4.2 demonstrates how the Delphi technique was used by engineers at the National Aeronautics and Space Administration (NASA). This design team was trying to improve the external insulation of an igniter case protecting O-ring seals from erosion during liftoff of a rocket. First, they assembled a group of experts and asked them to identify possible causes for the erosion they were detecting. The possible causes they identified are depicted in the row of boxes beneath the line with an arrow. This type of diagram is called a fault tree because it depicts a problem with possible causes (faults) branching off it. With the possible causes listed, the experts were then asked to determine which was the most likely cause. They used the following rating scale to make anonymous judgments:

Rating	*Description*	*Probability of being the cause of erosion*
5	Infrequent	0.1 (1 of 10)
4	Remote	0.01 (1 of 100)
3	Improbable	0.001 (1 of 1,000)
2	Very Improbable	0.0001 (1 of 1,000,000)
1	Almost nil	0.00001 (1 of 1,000,000,000)

An average rating was calculated for each possible cause. You see these averages listed in the rectangles below each possible cause. The fault tree (shown in Figure 4.2) was then distributed to the experts and they were asked to indicate whether they agreed or disagreed with the rankings that resulted from the ratings.

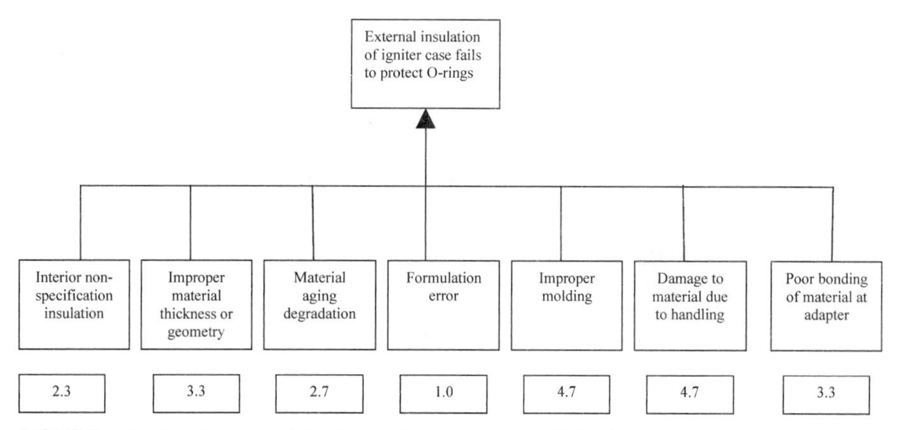

FIGURE 4.2 Application of the Delphi technique by NASA engineers.

They repeated the process a few more times until consensus was reached. When finished, the design team had a prioritized list of possible causes, which they used to allocate resources for further investigation. In this case the most probable causes were deemed to be improper molding and damage to material or threads.

A variation of the Delphi technique is the nominal group technique, which does not require anonymity. Here are the steps:

1. Identify your panel members and immediately ask them to generate ideas, solutions, or comments on the specific question. It helps if you have someone serve as a facilitator during this process. He or she is responsible for presenting the problem and the instructions to the rest of the team.
2. The team begins by quietly generating ideas for 5 to 15 minutes. You have several options here. They can share ideas verbally (as in brainstorming). Just make sure that there is no discussion/evaluation of ideas and that no one leaves until everyone is finished. Alternatively, each person can write down his or her ideas on paper. You can provide people with note cards and tell them to use one note card for each idea.
3. Next, collect the ideas and list them one by one on a computer projection screen, blackboard, or flipchart. If you have used note cards, you can collect them and post them on a wall. They should not be listed/posted in any particular order; however, to save time the facilitator should eliminate duplicate ideas.
4. Depending on the number ideas you have generated, you can ask each contributor to explain his or her concepts as you post, type, or write it. However, do not allow any evaluation or arguments. Discussion should be limited to clarification of meaning only.
5. After all solutions have been listed, ask each expert to nominally rank or rate them. You can have people walk around the room and place/write their ranking or rating next to each idea.
6. Once ideas have been assessed by all group members, you can obtain average rankings/ratings and thereby establish a prioritized list of your ideas. In other words, you have a nominal consensus.

Like brainstorming, this technique is a great way to generate many ideas and solutions in a short time. On the other hand, the limited discussion implied by the process may prevent people from fully understanding one another's ideas before they vote on them.

Another example from NASA depicts how the technique can be used. In this case, engineers were trying to develop a prioritized list of concerns they needed to address when designing chemical replacement technologies that were being mandated by law. A group of appropriate experts was assembled and they were first asked to generate as many concerns as they could think of. When they were done, a facilitator reviewed them and grouped them into the categories shown in Figure 4.3. Because so many concerns were generated, the facilitator assigned numbers to the concerns within each category. These are just tracking numbers and they are in parentheses to the left of each concern. Next, time was allocated to make sure everyone shared a common understanding of each concern/idea that was listed. No debate was allowed, just clarification. The group then voted on the importance of each concern. In this case, the team used a scale of 1 (least critical) to 20 (most critical). The votes were averaged to determine an overall importance level or weighting for each concern. These average ratings are listed to the right of each concern. The two concerns with the highest weightings were base material compatibility (item 6 under chemical concerns) and manpower dollars (item 1 under cost concerns). Can you identify the concerns with the lowest weightings?

4.2.4 TRIZ

TRIZ is the acronym for a Russian term that roughly translates to "theory of inventive problem solving." Genrich Altshuller, a Russian scientist, initially developed the concept in 1946. He began with the hypothesis that there are universal principles of invention serving as the basis for creative innovation across all scientific fields. He reasoned that, if these principles could be codified and taught, it would be possible to make innovation more predictable.To test his hypothesis, Altshuller reviewed over 200,000 patents submitted in what was, at the time, the Soviet Union (Russia).

Technical Contradictions Specifically, Altshuller looked at patents for solutions to what he called inventive problems. For a problem to be considered inventive it had to pose at least one contradiction. He defined a contradiction as a situation in which an attempt to improve on or meet one parameter of a design detracts from the design's ability to meet another. A contradiction between speed and sturdiness is one example. For instance, if we want to design a vehicle to be sturdy that typically means it will weigh more. But what if we want that same vehicle also to be fast? More weight generally means less speed and vice versa. TRIZ researchers have identified a total of 39 design parameters, each of which potentially could be in contradiction with one another. Although we will not list them all in this text, some additional examples include force, tension/pressure, temperature, power, brightness, adaptability, and repairability. The first step in using the TRIZ technique is to identify which design parameters are in contradiction with one another. In other words, redefine the problem in terms of the contradictions it poses.

Chemical Concerns	Avg. Weight
(1) Number of sources	7
(2) Limits of resources	7
(3) Availability	14
(4) Stability	15
(5) Drying ability	14
(6) Base material compatibility	17
(7) Toxicity	13
(8) Flash point	13
(9) Ease of maintenance	8
(10) Historical database	9
(11) Desirable reactivity	13
(12) Undesirable reactivity	13
(13) Lot-to-lot variability	11
(14) Shelf life	9
(15) Bond line thickness	7
Process Concerns	
(1) Contaminants removed	15
(2) Process steps	9
(3) Parts processed at one time	7
(4) Required surface protection	12
(5) Bondline thickness	7
(6) Process interaction	9
(7) Bondline strength	9
(8) Operator sensitivity	12
(9) Lot-to-lot variability	11
(10) General cleaning ability	13
(11) Surface requirements	14
(12) Possible stress crack	14
(13) Life of process part	14
(14) Damage due to process	13
Regulatory Concerns	
(1) OSHA requirements	13
(2) State/Local env. laws	14
(3) Federal env. requirements	14
(4) Future federal requirements	15
Safety Concerns	
(1) Worker exposure limits	12
(2) Spill response plans	13
(3) Fire response plans	14
(4) Explosion response plans	16

Environmental Concerns	Avg. Weight
(1) Clean air monitoring	12
(2) Pollution prevention	12
(3) Toxic emissions	15
(4) Emissions control	12
(5) Ozone depletor potential	15
(6) Chemical storage availability	10
(7) Ingredient recycling	10
(8) Hazardous waste management	12
Cost Concerns (dollars)	
(1) Manpower	17
(2) Facilities	15
(3) Materials	14
(4) Chemical	16
(5) Other hardware	14
(6) Contracts	12
(7) Specification changes	13
(8) Specification verification	13
(9) Change of drawings	11
(10) Procedure development	12
(11) Emissions control equip.	15
(12) Emissions control testing	12
Federal, State & Local Scheduling Concerns	
(1) Research	9
(2) Trade studies	8
(3) Modification in planning	9
(4) Specification documentation	10
(5) Requirements documentation	11
(6) Drawing/design changes	8
(7) Production time	11
(8) Testing	14
(9) Vendor selection certification	12
Present Program Schedule	
(1) Research	10
(2) Trade studies	11
(3) Modification in planning	10
(4) Specification documentation	11
(5) Requirements documentation	11
(6) Drawing/design changes	10
(7) Production time	11
(8) Testing	12
(9) Vendor selection & certification	11

FIGURE 4.3 Application of the nominal group technique at NASA.

Inventive Principles Once the problem has been redifined in terms of its contradictions we can then make use of another aspect of TRIZ. Altshuller determined that many of the same principles could be used to solve problems across an array of scientific fields and engineering disciplines. Although you may be working on a problem in one engineering discipline, chances are pretty good that an analogous problem has or is being considered within another discipline.

TABLE 4.4 Excerpted Portion of a TRIZ Contradiction Table

Undesired Result (conflict) / Feature to Change	1 Weight of moving object	2 Weight of nonmoving object	3 Length of moving object	4 Length of nonmoving object	5 Area of moving object
1 Weight of moving object			15, 8, 29, 34		29, 17, 38, 34
2 Weight of non-moving object			10, 1, 29, 35		
3 Length of moving object	8, 15, 29, 34				15, 17, 4
9 Speed	2, 28, 13, 38		13, 14, 8		29, 30, 34
10 Force	8, 1, 37, 18	18, 13, 1, 28	17, 19, 9, 36	28, 10	19, 10, 15

For example, the same scientific principles used to split gems are also used to extract seeds from bell peppers so that they may be canned. In both cases, the goal is to break something apart without shattering it. The technique to do so involves placing the item in an airtight container, gradually applying increased pressure, and then dropping that pressure quickly. The underlying principle is that a sudden drop in pressure creates an explosion, which splits the object along a more or less uniform fracture line. We do not mean to imply, of course, that the technique is identical in both cases. TRIZ may help to generate the idea but the hard work of determining the correct conditions and materials still must be done. In the case of diamond cracking, the 8 atmospheres of pressure used for bell peppers would not work. Experiments and tests must be done to ensure appropriate application of a principle to a given problem. The point is, however, that by systematically codifying which principles can help us solve given combinations of contradicting parameters, we can speed up the idea generation process instead of leaving it to trial and error. For instance, the process for splitting and canning peppers was patented in 1968. The patent given for crystal splitting was not issued until almost 20 years later.[1]

TRIZ provides this codification of principles relative to contradictions in the form of a contradiction table.[2] Table 4.4 depicts portions of this table. The complete table lists all 39 parameters along both its vertical and horizontal axes, the numbers in the smaller boxes, where the vertical and horizontal axes intersect, refer to the design principles that have been used in the past to help solve the contradiction posed by the two intersecting contradictions.

The table has been built by reviewing and codifying what is, at this point, over 2 million patents. As you can see, even from this excerpted portion of the table, not all contradictions have been solved. However, by reviewing the table, engineers and scientists can obtain some ideas about where to start in terms of looking for solutions.

[1] J. Terninko, A. Zusman, and B. Zlotin. *Systematic Innovation: An introduction to TRIZ,* St. Lucie Press, 1998.

[2] *Ideation Methodology: The Training Manual* (4th ed.), Ideation International Inc., Southfield, MI, 1996.

A total of 40 principles have been identified that transcend engineering fields and scientific disciplines. As was the case with regard to the contradictions, it is beyond the scope of this text to detail them all, but we do want to give you some examples.

One TRIZ design principle is the concept of segmentation or what is sometimes called division. This technique involves dividing up an object that it usually perceived as one whole. Multistage rockets are an example of the application of this principle. For a rocket to be sent into space it must be able to overcome gravity. Doing so requires a very high rate of speed (escape velocity). However, to reach this speed a rocket equires a great amount of fule. The large amount of fuel, of course, increases the weight, which in turn makes it impossible for the rocket to reach escape velocity.

Solving the problem involved breaking the rocket into several smaller sub-rockets, each of which was essentially a small fuel tank. As its fuel is expended, the subrocket, falls away, which means that each subsequent rocket applies its force on a smaller mass and accelerates what remains until the satellite or spacecraft has sufficiently escaped the earth's gravitational force.

A more day-to day example of segmentation is a car radio with removable parts that prevent theft. As in the case of the rocket, segmentation helps us overcome contradicting design parameters, which in this case are reliable usage (enabling the radio to work when we want it to) but at the same time decreasing the likelihood of a harmful side effect (theft).[1]

Another TRIZ principle is the concept of multiplication. The principle suggests deriving solutions by duplicating some component that already is inherent to the problem. Consider vaccinations against viruses. Typically, they are altered versions of the very virus they are trying to prevent. The original virus is modified so that it is harmless to most people. This weakened duplicate is then used as a catalyst that prompt our immune systems to create a resistance.

A second example of the multiplication principle may be found in efforts to design systems to prevent major earthquakes. Earthquakes occur because the earth's outer shell is broken into 11 huge solid plates floating on a layer of molten rock. Earthquakes start along the boundaries where these plates meet. As they float about, these plates are constantly jostling each other. However, in spite of all this jostling, the plates don't just slide past each other. More typically, they grind together much like two large pieces of sandpaper. In many instances they get stuck for hundreds of years. Because the plates are still separate entities wanting to move in their own directions, pressure builds up at one or more points where they are stuck. Eventually the pressure reaches a point where it abruptly bursts the two plates apart again and causes what we know as an earthquake.[2] Greater amounts of built-up pressure lead to earthquakes with greater force. Geologists reason that if they can keep the pressure from building up they can prevent earthquakes. One solution is to alleviate the pressure gradually by inducing a series of small earthquakes. Currently, this solution is far from being perfected but many scientists are hopeful that it holds some promise.

[1]R. Horowitz. *Creative Thinking: Professional Version Training Manual*, BCI Software, Columbus, OH, 1999.

[2]D.Thompson, "Can We Save California?" *Time Magazine*, Vol. 155(14), April 10, 2000.

Today, numerous organizations are using TRIZ to solve design problems and develop new technology strategies. These organizations iclude Ford, Motorola, Proctor & Gamble, 3M, and Siemens among others. As we mentioned earlier, TRIZ principles to not necessarily give you the answer but they are likely to point you in the right direction. In addition, a complete understanding of TRIZ requires a lengthier discussion than we have provided here. Nonetheless, even this limited discussion offers some general guidelines that can be of immediate value in formulating potential solutions for your own design projects. four specific points to take away are as follows:

- Design solutions may be found by looking across scientific fields of study and engineering disciplines. Be open to the fact that people working in different fields may have dealt with analogous problems.
- An effective design solution does not always include adding something new to the problem. Be open to ways you might be able to reorganize what is already there and/or use it differently.
- It helps to define your design problem in terms of the contradictions it poses.
- There are fundamental principles that have helped guide people to design solutions and these principles transcend disciplines and fields of study.

4.3 EVALUATING AND ANALYZING ALTERNATIVES

By the time you have finished using techniques like brainstorming and lateral thinking, odds are pretty good that you have identified quite a few alternatives. Of course you will not have the time, money, or other resources to try them all. You will have to systematically choose among them and determine which have the greatest likelihood of meeting your design problem's functional requirements in light of its constraints. This is what we mean by evaluating and analyzing alternatives.

In the design process, a variety of factors must be included and considered when alternatives are being evaluated and analyzed. Some of these are cost, strength, weight, speed, time, rigidity or flexibility, appearance, reliability, effect on the environment, corrosion resistance, safety factors, and a variety of others depending on what you are designing. In many cases, the decisions about components will be made separately, and then subassemblies or total assemblies have to be analyzed again so the combination of factors can be considered. A typical approach is to look at the apparent options and make a table that includes the options in a column on the left with a row of the factors across the top.

4.3.1 Systematic Decision Grids

A decision grid is one useful technique for evaluating your designs in relationship to specific criteria such as those described in the preceeding paragraph. To use this technique, put together a small table that lists the factors that are important for your design or some subcomponent of the design (see Figure 4.4). These factors typically

Motor-Gear Train Comb	Cost	Torque	Speed	Power	Noise
1					
2					
3					
4					
5					

FIGURE 4.4 Systematic decision grid.

are the functional requirements and parameters for your design. For instance, if you were creating a decision grid to evaluate ideas for the design of the motor gear train subsystem of the autonomous robot it may include the factors indicated in Figure 4.4.

Although Figure 4.4 looks pretty good, your team might have had the foresight to add some other factors such as reliability, if you have such information, or weight or ease of attachment to the chassis. Once you have created the grid, your team can discuss the pros and cons of each design alternative objectively in terms of the factors (functional requirements and design parameters) you identified. You can fill in the matrix by placing a plus sign in the column if the alternative has a beneficial impact or a minus sign if the alternative does not meet the requirement. Or you can rank each alternative in relation to each factor. For example, assign a 1 to the alternative that best meets the requirements of the factor, a 2 to the next best alternative, and so on. Assuming the factors are equally important, you can then choose the alternative with the most pluses or, if you used the ranking system, the alternative with the lowest score.

With regard to power, you should realize that batteries do not provide a constant source of voltage over the "life" of a single charge. The rate of decline varies by the type of battery, temperature, and other factors. The teams find that it makes sense to measure the available power over time during practice runs so that they know they will have sufficient power during the competition.

Let us look at the chassis for the robot. It is possible to construct the chassis from a variety of components: Erector sets, Legos, PVC pipe and fittings (and adhesive), wood, metal, PVC sheet, and structural shapes. Again, after your team has done the brainstorming about shape, size, power train, pickup mechanisms, sensor mounting, and so forth, you need to make a table for the chassis materials and components, including the factors for making your decision. The list might include cost, ease of assembly, ease of modification, ease of attaching sensors, pickup mechanisms, and power trains.

Once the initial decisions on power train and chassis are made, it probably is wise to check on the total weight of the chassis, power train, and controller to see whether you have enough power from your power train.

If you don't have enough power or speed you will have to make some decisions about whether to change, and this can include whether you have enough time to make the change to meet the project guidelines and deadlines. One team had a reliable robot that was not as fast as they would have liked. It just barely met the time requirement. They changed the power setting and found that the robot was not as reliable at the higher power level. They chose to stay with the design setting just for the reliability and found that they were very competitive with faster but less reliable robots.

4.3.2 Force-Field Analysis

A force-field analysis is another good tool for evaluation and is particularly useful after a rich brainstorming session when you are trying to achieve consensus. Complete, accurate records are essential. Here are the steps to follow to do a simple force-field analysis:

1. Identify all forces as positive or negative in terms of how they affect the project.
2. Assign a value or numerical weight to each force (for instance, 1–10, with 1 being the weakest force and 10 being the stronger force).
3. Add up the positives and negatives.

The more positive the total, the more likely the solution is correct. The more negative the total, the more likely the solution is unworkable. One solution will emerge as meeting the least resistance. A strategy is then developed to enhance the positive forces and reduce the negative, either at the end of brainstorming or during evaluation. If good records are kept and they are analyzed accurately, a force-field analysis can help you develop a good set of lessons learned for the next project.

For instance, let's say you are considering whether to use molded plastic as the material for designing a boat. Your force-field analysis might look like Figure 4.5a.

As another example, your team meets to discuss whether to work on the design project this Saturday morning (Figure 4.5b). Identify the positive and negative forces that may affect your decision. Positive forces clearly, but not overwhelmingly, outweigh negative ones. Now your task as team leader is to lessen the magnitude of negative forces when you meet Saturday morning. For example, you can start at 10 a.m. instead of at 7:30 a.m., schedule a final review the night before the project is due, and work smart so you do not have to make excuses for a late or shoddy project.

4.3.3 Making Estimates

As you begin evaluating your design ideas, chances are good that you will also realize you do not necessarily know everything you need to know in order to fully evaluate your options. Remember this is still a relatively early point in your overall design process. Although you may have a general idea of what your design requirements

a

Positive Forces		Restraining Forces	
Light weight	7	Can crack or tear	6
Inexpensive	6	More costly to repair	6
Noncorrosive	7	Not as durable	5
Total Positive	20	Total Negative	17

b

Positive Forces		Negative Forces	
Deadline looms	8	We are tired	8
Get it over with	9	Wait until last minute	6
Rest up Sunday	9	Make up excuses	4
Total Positive	26	Total Negative	18

FIGURE 4.5 Force-field analyse.

are, you still may be lacking some of the specifics. For instance, you may know that light weight is one functional requirement but what does that really mean? How much should a bicycle weigh? How much will a robot weigh? How much will a robotic arm weigh? You will not always have enough information to know precisely, so an estimate provides a starting point. As you progress through the design process you can recalculate things you need to know and get closer to the answer you need to make a good decision.

How do experienced engineers make estimates? They do this on the basis of the background and knowledge they have acquired when they did previous designs. Surprisingly, you and your teammates already have a fair amount of information you can use. You also need to use your common sense about the information. For example, when you are concerned about weight, you know a bicycle does not weigh 200 pounds, nor does it weigh just 2 pounds. What does a bicycle weigh? When you are concerned about electrical power you can think about a number of things you probably know already. The voltage in the wall receptacle normally is 110 volts AC. The battery in your smoke alarm is 9 volts. The AA batteries are 1.5 volts. A computer's power supply takes 110 volts AC and about 3 amps. The voltage used in the components inside is normally 5 volts or 3.3 volts and the current is in milliamperes. Water pressure in a residential area or office complex is normally about 60 pounds per square inch (psi). The atmospheric pressure is 14.7 psi at sea level on a standard day. A car weighs 3,000 to 4,000 pounds or 1.5 to 2.0 tons. It may be worthwhile for you to make a list of known things to make estimates for your unknown design.

Part of the information also can be obtained by looking at catalogs, the Web, and textbooks. As you gather the information, make a copy, and put it in your project notebook; carefully note when, where, and from whom you got the information.

Still other information you will need to make decisions has to come from calculations that you and your team perform based on information from your classes, textbooks, reference books, handbooks, and other such sources. Each time you perform or attempt to perform a calculation as part of the analysis, make sure a copy of the work is placed in your project notebook. Even the calculations that did not provide what you wanted can be valuable so that someone who comes after you knows not to go down that road.

As an example, consider the fact that students working on the autonomous robot know speed is an important factor. The robots must move fast enough to complete the course in the specified time. But how fast must it be? The desired speed can be determined by measuring the path the robots must follow and dividing it by the time allowed for a run. To this distance and time consideration the students should add for contingency factors. The robot may run into something unexpected, such as another robot or an obstacle, and extra time will be required. Thus, the speed required is greater than the base level speed to travel the course with no interference.

The robots also must be able to climb a ramp, so power is a consideration. To solve this problem, the students must sketch a view of the robot going up the ramp, draw a free body diagram with appropriate forces, and complete the analysis. Things become a little more complicated at this point. The students have not yet built the robot so they don't know the weight. This points to a technique that is important in doing analysis for design. When setting up and solving equations to provide answers, write the equations in terms of variables that may change. Estimates of factors such as weight can be made initially, and as the process continues and more components are added to the design, the team will get closer and closer to the real weight. It also is possible that the robot may not approach a ramp at exactly 90 degrees and may attempt to go up the ramp at an angle. Is this be easier or harder? Again, it makes sense to determine the torque required from the motor or motors and add in a little extra to overcome things such as friction and the amount of power available from the batteries.

Motor systems come in a variety of configurations. They can be purchased as a simple motor, in which case a gear train or transmission has to be designed to go with the motor. It also is possible to purchase gearhead motors that have an internal gear train built in or bolted on. Again, some analysis is required to determine whether the motor/gear train combination will provide enough speed and power to make the robot perform well.

The electrical noise a motor makes and the amount of power it draws are factors that are very important when the robot's controller and the drive motors draw power from the same battery pack. Most controllers (small computers) require a set, constant voltage to maintain the contents of memory. When the robot starts up the ramp, extra power is be required over and above the power to move on level ground. If the power drawn from the batteries is too high, the voltage may drop temporarily, and the controller can lose its memory. Furthermore, electrical noise from the motor could possibly interfere with the controller electronics. Such difficulties can be partly alleviated by connecting the motor to the controller with a variety of components in the circuit to isolate the noise and also to prevent the voltage drop. However, these components cost additional money and time. Obviously, it would be nice to find a motor without these interfering characteristics and have it connected to a reliable gear train. The combination would provide the power and speed to be competitive. And, by the way, the cost must be within budget.

4.4 SELECTING A SOLUTION

Systematic evaluation and analysis will eventually lead you to a potential design solution. It is the "output" or results from your evaluation and analysis that become the basis for selecting a solution or solutions to be pursued further.

It is important to realize that the overall project schedule and length of time to achieve a solution should be part of the process of evaluating, analyzing, and selecting a solution. An innovative new product may not be successful if it cannot be made available in time to meet the market demand. Such products that fail to be successful in this way are said to have missed their "market window." Thus, as progress is being made on selecting a solution, other elements of the design process must be brought along concurrently. Important among these elements is the design schedule and the project proposal. In addition to the techniques for generating and evaluating design alternatives, we include some discussion of several skills and tools for developing project schedules and preparing proposals in the next section.

Keep in mind that, whether they are working on student engineering design projects as a first-year student or on real-world projects, the best engineers seek simple solutions. There are a variety of reasons for this, but an illustration or two might help. When Toyota was designing and developing their 1998 Camry, to reduce the production cost and list price over the 1997 model, they first sought better understanding of their design and then began modifying individual components. One excellent example was the design of the bumper. In the earlier model, the bumper assembly had more than ten components. The new model had a bumper with three components. This reduced the total component cost, and, more importantly, it reduced the assembly cost for the new car. The design production team for the Corvette used a similar approach to provide a better product at a lower cost.

Design simplicity can take several forms. In the PC design, a very inexpensive model can be produced by putting all the components on a single printed circuit board. The disk controllers, the video electronics, the keyboard, and mouse interface all on one board significantly reduces the cost of a unit. However, the technology is moving so rapidly, that, for example, video technology can quickly outstrip the built-in capability and the owner is "stuck" with a unit that works perfectly well with the original software but is incapable of being upgraded with a new video board.

Therefore, we have the initial cost and the "keeping up to date" or upgrade cost to consider. This is where the automotive industry and PC industries vary rather widely. Most cars are built and never modified throughout their lifetime. We simply use them until it is too expensive to buy and install replacement parts. PC technology is moving so rapidly and providing such productivity gains (or at least better output) that we are compelled to move quickly from one model to the next. Civil engineers may find that a bridge designed in the 1860s or 1880s still may function quite adequately, although a replacement probably will not be designed in the same way as the original.

CHAPTER REVIEW

The formulating solutions phase means that you will be identifying various design alternatives, defining the parameters of design, evaluating those alternatives, and then seeking a potential solution. You should realize by now that obtaining a potential solution is a systematic process requiring you to carefully consider multiple alternatives and ultimately select among them. Keep in mind that identifying solutions and evaluating them are related but distinct activities. Push yourself to consider several design alternatives before beginning the process of evaluating their merits. Using techniques like brainstorming and lateral thinking can help ensure that you and your design team consider a range of options. When evaluating your ideas, other tools like a decision-matrix and force-field analysis can help structure your thinking. You also need to provide a clear rationale for your eventual solution and these tools can help you both clarify and communicate your thought process. Remember also that successful solutions do not necessarily have to be wholly original ideas. Incremental innovation and application of concepts from other engineering disciplines are a cornerstone of good engineering design. Finally, keep in mind that engineers (especially at this early stage of a design process) rarely have *all* the information they would like in order to make decisions. Therefore, appreciate the value of making well-reasoned estimates and inferences.

REVIEW QUESTIONS

1. What are the typical steps in the formulating solutions phase of the design process? How are they related? How are they distinct from one another?

2. What are the four basic rules or guidelines for productive brainstorming? How is lateral thinking different from brainstorming?

3. Use the random word list on page 83 as a prompt to generate design solutions to your own design problem, assigned by your instructor. (*If you have not been assigned a specific problem apply the technique to the problem introduced in Section 4.2.2—a device for keeping a book dry when in the bathtub.*)

4. What is the difference between innovation and origination?

5. What are the differences between vertical thinking and lateral thinking? How do they complement one another? Which decision-making tools discussed in this section best reflect vertical thinking? Why?

6. Explain how TRIZ can help engineers formulate design solutions. Do you think TRIZ has anything in common with lateral thinking? Why or why not?

DORM ROOM DESIGN PROBLEMS FOR CHAPTER 4: FORMULATING SOLUTIONS

Assignment 7 (decision making)

Now that your team has a thorough understanding of the problem, it is time to do some brainstorming about how to solve the problem as a whole and all the little problems that are part of the larger one. Remember that when you are brainstorming you must accept all ideas and build on them no matter how ridiculous. Find a place where your team is isolated enough to be able to hear and see each other. Sit around a table or gather your chairs together in a circle. If you have a small team (about 4), each person should have a pad of paper and pencil to write things down. If you have a large team (about 8), assign two of the people to be recorders. Don't worry about organization of ideas. That will come later. Plan to spend 10–15 minutes of concentrated activity where your minds are focused on the problem at hand. Try using some lateral thinking techniques as well. Document the techniques you try and the ideas you generate.

Assignment 8 (decision making)

Take the list(s) of ideas and sort them into categories. For example, there will be ideas associated with the bed, the desk, the place for the computer, the bookshelves, clothes storage, study chair, telephone, network lines, etc. Your task is to put together 6 to 12 solutions for the problem. This can be various combinations of the ideas to solve the smaller parts of the problem. Draw sketches of about 6 good ideas for solving the entire problem. Draw these on engineering problem paper or other grid paper. Document the idea with the appropriate notes on the sketch and lists of critical items on a second or third sheet. The documentation should include who drew the sketch, the date, the class, etc.

Assignment 9 (decision making)

Now that your team has 6 or more good ideas for solving the problem, lay them out on the desk or pin them on a bulletin board and decide what the design parameters should be. Go back to your interviews with your fellow students to help make up the list of parameters. Do your designs take into account all the parameters that were generated in your interviews? If your sketches left something out, go back and see if they can be updated. Remember that your budget is limited and you want the best design that meets the need. Make a table of the ideas in the left column and the parameters across the top and fill in the table with checks or brief notes to see whether you have taken care of all items.

BIBLIOGRAPHY

DEBONO, E. *New Think: The Use of Lateral Thinking in the Generation of New Ideas*. Basic Books, New York, 1967.

DEBONO, E. *Lateral Thinking: Creativity Step by Step*. Harper-Perennial, New York, 1990.

FENTIMAN, A. *Team Design Projects for Beginning Engineering Students,* Technical Report ETM-10-05-958. Gateway Engineering Education Coalition, Philadelphia, 1997.

HOROWITZ, R. *Creative Thinking Professional Version Training Manual.* BCI Software, Columbus, OH, 1999.

Ideation Methodology: The Training Manual (4th ed.). Ideation International Inc., Southfield, MI, 1996.

OSBORNE, A. *Applied Imagination*, Scribner's, New York, 1957.

SCHEFTER, J. *All Corvettes Are Red.* Pocket Books, New York, 1996.

Sunrayce, General Motors, U.S. Department of Energy and Electronic Data Systems. Web site (www.sunrayce.org), 1999.

TERNINKO, J. ZUSMAN, A., and ZLOTIN, B. *Systematic Innovation: An Introduction to TRIZ.* St. Lucie Press, 1998.

THOMPSON, D. "Can We Save California?" Time magazine Vol. 155(14), April 10, 2000.

CHAPTER 5

FORMULATING SOLUTIONS: PROJECT AND PEOPLE SKILLS

In the preceding chapter we outlined *what* you must do to begin formulating solutions. We also discussed some key decision-making tools that will help you get the work done. In this chapter, we focus on relevant project management, communication, and collaboration skills. We introduce specific aspects of each of these skill areas that are best applicable to this second phase of the engineering design process. The shaded boxes in Table 5.1 list the topics we will cover.

You have already added a number of very useful skills to your "tool kit" as a result of completing design Phase 1, and you will have the opportunity to apply some of these skills as you formulate solutions. But we want to add several new tools and tactics that are specifically well suited to this phase of the process. In terms of project management skills, this second phase of the design process describes the importance of maintaining a record of the information and facts gathered about the current design problem and previous work related to the problem. Some management tools for preparing a design schedule are included as well. The new communication skills introduced in this phase directly support the project management skills by providing guidelines for sharing and preserving the project information. Finally, we offer some insight about how to keep the team working together as a unit in the presentation of some new collaboration skills. Table 5.2 relates specific skills to each design step discussed in the preceding chapter. It provides an overview of when you will be most likely to benefit from using each skill and tool that is described.

5.1 PROJECT MANAGEMENT

During the course of formulating solutions you will also begin to a get a clearer sense of all the work and activities you need to perform. This is the point in your project where you will want to finalize your work schedules and plans. In this section we discuss techniques for project planning and also describe methods you will want to use for monitoring your progress.

TABLE 5.1 Overview of Phase 2: Formulating Solutions

Steps for formulating solutions	Skills and tools for formulating solutions			
	Decision making	Project management	Communication	Collaboration
5. Defining design parameters 6. Identifying alternatives 7. Evaluating and analyzing alternatives 8. Selecting a solution	• Innovation versus origination • Considering external factors • Brainstorming • Nominal group and Delphi techniques • Lateral thinking • Systematic decision grids • Force-field analyses • Making estimates	• Preparing and using Gantt charts • PERT/CPM techniques • Establishing and maintaining records	• Sharing the data gathered • Writing proposals • Preparing bibliographies	• Ensuring open participation • Reaching consensus and building commitment • Managing conflict • Avoiding groupthink

5.1.1 Preparing and Using Gantt Charts[1]

A good design proposal typically includes a timeline that shows how you will progress toward completing your project. A timeline is simply a proposed schedule based on your best estimate of what the necessary tasks are, when you will do them, and how long they will take to complete.

Scheduling is an important part of any design project. It arises in two different contexts. The schedule lays out expected progress in conducting the design itself, and the schedule is used for realizing the design (actually building it).

One of the simplest yet most informative ways to present a schedule is with a Gantt chart, named after its developer. The Gantt chart is simply a list of all the tasks necessary to complete a project arranged along a timeline. The timeline shows the scheduled start of each task and its anticipated completion by horizontal bars starting when the task is to begin and ending when it is supposed to be completed.

Several elements are necessary to successfully constructing a Gantt chart for a design project:

- Identifying the tasks needed to complete the design,
- Estimating the time required to complete each task,
- Determining whether there is a sequence in which the tasks must be scheduled, and
- Identifying deadlines.

[1]R. Weggel, V. Arms, M. Makufka, and J. Mitchell. *Engineering Design for Freshmen*, Technical Report #ETM-1058. Gateway Engineering Education Coalition, Philadelphia, 1998.

TABLE 5.2 Relationship of Skills and Tools to Each Step in Phase 2 of the Design Process: Formulating Solutions

	Phase 2: Design steps			
Skills and tools	**Defining design parameters**	**Identifying alternatives**	**Evaluating and analyzing alternatives**	**Selecting a solution**
Project management (3.2.2)				
▪ GANTT			X	X
▪ PERT			X	X
▪ Record keeping	X	X	X	X
Communication (3.2.3)				
▪ Writing the proposal	X	X	X	X
▪ Preparing a bibliography	X	X	X	X
Collaboration (3.2.4)				
▪ Ensuring participation	X	X	X	X
▪ Reaching consensus			X	X
▪ Managing conflict	X	X	X	X
▪ Avoiding groupthink			X	X

Identifying Tasks Carefully think about all the tasks necessary to complete your design and list them. For example, a generic list (in no particular order) might include the following:

- Conduct literature review (library research)
- Make personal contacts (information gathering)
- Prepare proposal (write)
- Obtain necessary data
- Analyze data
- Identify alternatives
- Formulate preliminary design
- Evaluate alternatives
- Prepare final report
- Identify design topic
- Prepare final design
- Assemble design team
- Prepare presentation
- Rehearse presentation
- Revise presentation
- Make presentation

- Prepare draft final report
- Revise final report
- Submit final report

Identifying the Time Sequence It is obvious in the preceding list that some tasks must be completed before others can begin. You cannot write your proposal until you have identified a design topic, for example. Similarly, you cannot rehearse your presentation until you have prepared it. It is also obvious that some tasks are really firm deadlines, such as "make presentation" and "submit final report." It is easy to locate these on the Gantt chart since your client, boss, or professor usually specifies them. They are occasionally identified as a large asterisk on the chart at the time when they are due. The remaining tasks can be arranged approximately in the order in which they must be logically started.

Estimated Task Duration Most of the tasks on the list require some time to accomplish. Estimating the amount of time necessary depends on experience. (Some engineers tend to be overly optimistic about how long it takes to do anything.) In the absence of experience, be conservative but realistic. Remember, in most student design projects you have nonnegotiable deadlines to meet. Anticipate problems and plan accordingly.

Constructing the Chart A typical Gantt chart is shown in Figure 5.1. The tasks are listed approximately in sequential order, and a horizontal timeline is constructed at the bottom. Deadlines are indicated with asterisks. Usually, open bars of appropriate length are used to indicate the duration of the task. (Open bars are used so they can be filled in to indicate progress.) The bar begins when the task is scheduled to start, and the bar ends when the task is expected to end. This is where judgment comes into play. Some tasks overlap and can be done simultaneously. After all, there are several team members to work concurrent tasks. Other events must await completion of an earlier task. For example, you can be writing some parts of the final report while you are completing the final design, but you can't revise your report until after it has been written and reviewed.

Using the Chart The Gantt chart is used in at least two main ways: as a proposed schedule (in your proposal) and as a progress chart. It can also serve to identify what resources are needed and when they are needed. For example, if an entry on the chart reads "analyze data," that implies the data will be available for analysis. You also might look at the chart to see when extra resources or personnel are needed to meet a deadline (e.g., for writing and preparing figures for the final report).

To ensure that you are on schedule, the chart should be referred to as your project progresses. Obviously, things change. You cannot have anticipated everything that occurs as you move through your design project. When things do change, you can update your schedule, but remember, your deadlines are usually fixed!

The chart also can serve as a record of progress. As a task is started and completed, the open bar is filled in and the actual start and finish dates are shown next to the scheduled start and completion times. (On future projects, this information can be used to gain a better estimate of how long it takes to complete various tasks.)

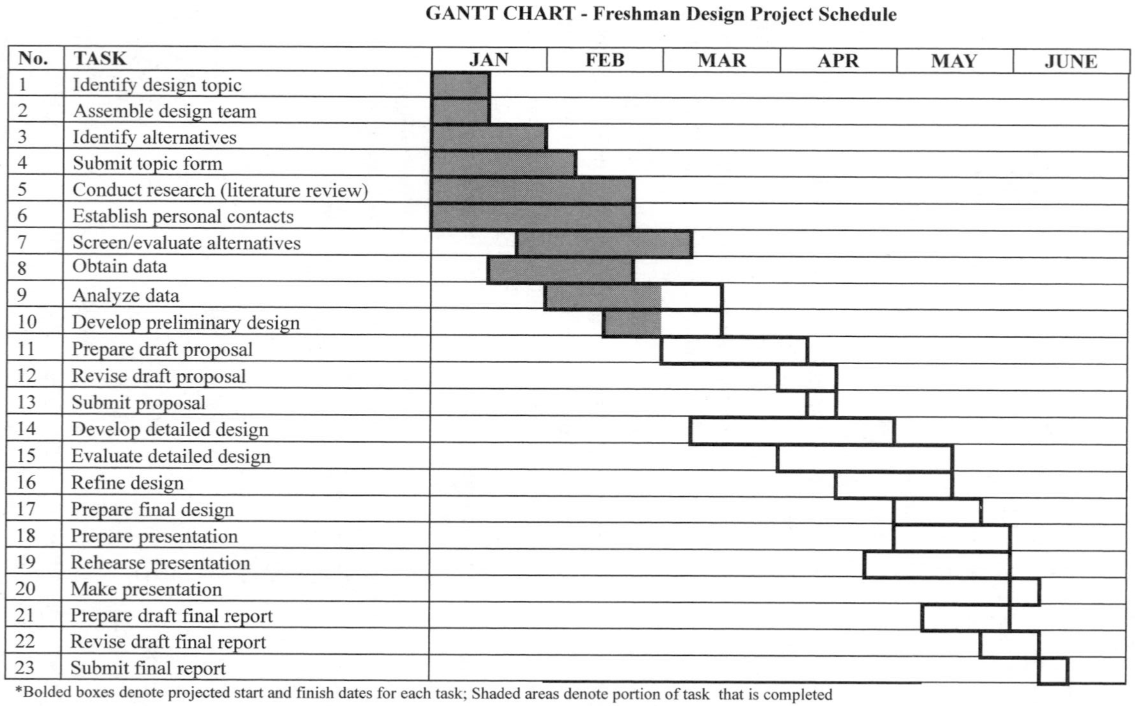

GANTT CHART - Freshman Design Project Schedule

No.	TASK	JAN	FEB	MAR	APR	MAY	JUNE
1	Identify design topic						
2	Assemble design team						
3	Identify alternatives						
4	Submit topic form						
5	Conduct research (literature review)						
6	Establish personal contacts						
7	Screen/evaluate alternatives						
8	Obtain data						
9	Analyze data						
10	Develop preliminary design						
11	Prepare draft proposal						
12	Revise draft proposal						
13	Submit proposal						
14	Develop detailed design						
15	Evaluate detailed design						
16	Refine design						
17	Prepare final design						
18	Prepare presentation						
19	Rehearse presentation						
20	Make presentation						
21	Prepare draft final report						
22	Revise draft final report						
23	Submit final report						

*Bolded boxes denote projected start and finish dates for each task; Shaded areas denote portion of task that is completed

FIGURE 5.1 Typical Gantt chart for freshman design project.

5.1.2 Program Evaluation and Review Technique and Critical Path Method

At the height of the Cold War, just as Sputnik was launched by the Soviets, the U.S. Government launched a number of massive engineering projects for which the simple planning tools like Gantt charts were inadequate. Booz-Allen, a management consulting firm, created the program evaluation and review technique (PERT) for the Polaris weapon system of the Navy. PERT soon caught on in the aerospace and research and development industries. At about the same time DuPont developed a technique they called critical path method (CPM). In theory there are subtle distinctions between PERT and CPM, but in practice they are essentially the same and we describe them here as such.

PERT/CPM is basically a tool used for complex planning and program control and is better suited for large programs than for small projects. The process is usually too complex for most entry-level student engineering projects. Nonetheless, you should at least be familiar with this technique because it is widely used in the engineering world.

Although your own projects may not be as massive an undertaking as designing the Polaris missile system, a basic understanding of PERT/CPM can help you coordinate the work of your design team. The main limitation of a Gantt chart is that it portrays project tasks as independent actions. Because it does not show the way tasks are interrelated, a Gantt chart cannot convey how schedule changes or time delays on one specific task might affect the overall project. This is where PERT/CPM is a valuable tool. PERT/CPM helps you prioritize work and shows at a glance what event comes first, next, and concurrently. As a result, it can help you recognize the critical consequences of schedule changes. It also helps you determine what tasks can be done in parallel and what should be done sequentially. Here are the steps for building a PERT/CPM chart.

1. *Prepare a work breakdown structure*. As when you prepare a Gantt chart, the first step is to identify and list all the tasks associated with the project, along with an estimate of the time required to complete each. This list is often referred to as a work breakdown structure (WBS). Figure 5.2 shows a WBS for a 25-week project to design and implement a computer system.
2. *Create an activity network and identify the critical path.* Once the tasks have been identified, the next step is to prepare a diagram showing how the tasks in the project are related to one another. This diagram, known as an *activity network*, depicts which tasks come first, which come second, and which can be done in parallel (see Figure 5.3). Once you have created the network, you need to identify the *critical path*. The critical path is the path of activities that takes the longest to complete. In our network, the critical path goes from Task 1 (system specification) to 2 (define data dictionary) to 4 (screen and report layouts) to 13 (select database manager) to 15 (develop backup/restore) to 20 (integrated testing) to 21 (train users).

   The overall time it takes to complete a project is the sum of the time required to complete each activity along the critical path. In our example, this

Task	Duration (Weeks)	Task	Duration (Weeks)
Start	0	12. Prepare Test Data	3
1. System Specification	3	13. Select Database Manager	3
2. Define Data Dictionary	3	14. Install Hardware	2
3. Hardware Specification	3	15. Develop Backup/Restore	6
4. Screen & Report Layouts	5	16. Program Screen and Reports	3
5. Select Terminals	2	17. Program Database Definition	3
6. Select Computer	2	18. Test Terminals	1
7. Design Test Procedure	3	19. Write Users Manual	2
8. Site Preparation	2	20. Integrated Testing	3
9. Computer Order & Delivery	5	21. Train Users	2
10. Terminal Order & Delivery	5	End	0
11. Write Terminal Interface	3		

FIGURE 5.2 Work-breakdown structure for designing and implementing a computer system.

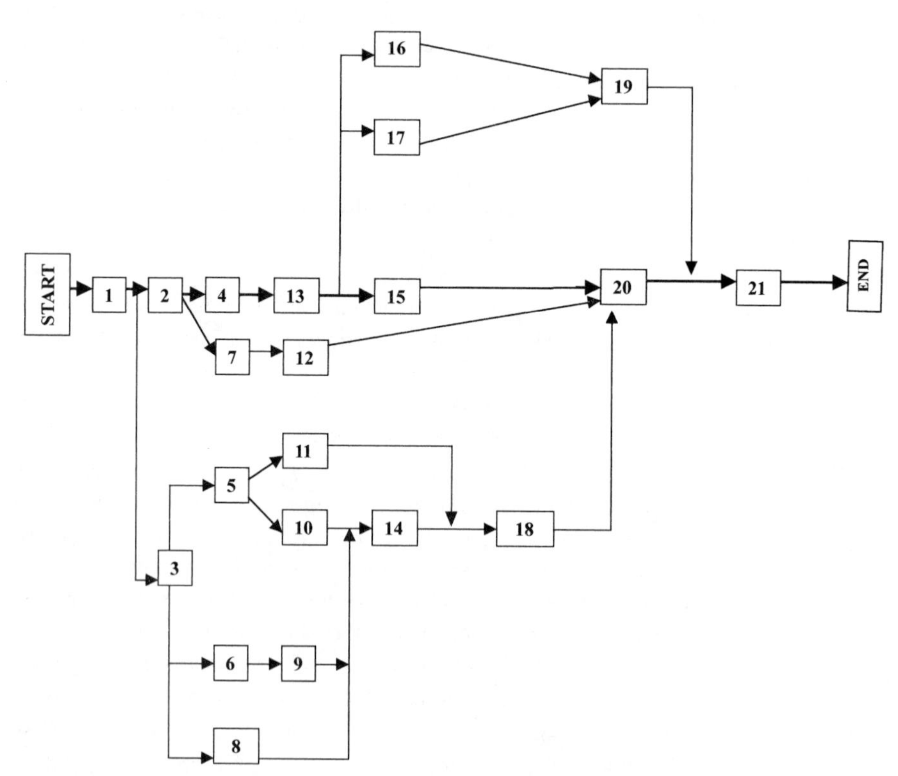

FIGURE 5.3 Activity network for designing and implementing a computer system.

sum is 25 weeks ($3 + 3 + 5 + 3 + 6 + 3 + 2 = 25$). You should notice that in addition to the critical path there are other activity paths. For instance another path goes from Task 1 to Task 3 to Task 5 to Task 10 to Task 14 to Task 18 to Task 20 to Task 21. The work along these other paths also must be completed but none of them will take as long to complete as the work along the critical path. Can you identify other paths in Figure 5.3?

Keep in mind that the amount and organization of paths in an activity network often depends on the number of resources you have available. Specifically, you cannot follow parallel paths if you do not have the people or material needed to work on them.

3. *Determine the earliest start time for each activity.* Having diagrammed relationships between tasks in your project, the next step is to calculate the earliest start times for each activity in the network. To calculate the earliest start times, move from right to left along your network. Do the critical path first and then the noncritical paths. In our example the early start times along the critical path are as follows: Task 1 starts at Week 0. It takes 3 weeks to complete (see the WBS in Table 5.2) so Task 2 begins at Week 3. The next step along the critical path is Task 4. It begins at Week 6 (because Task 2 takes 3 weeks to complete). Moving down the critical path, Task 13 must begin at Week 11, Task 15 at Week 14, followed by Task 20 at Week 20 and Task 21 at Week 23. Follow the same procedure for calculating early start times of the tasks along your noncritical paths.
4. *Calculate the latest possible start times for each activity.* Do this by moving from right to left along the activity network, beginning again with the critical path. You should notice that the latest start times for critical path activities are the same as the earliest start times (see Figure 5.4). You do, however, have some flexibility when it comes to tasks along the noncritical paths. For instance, another path goes as follows: starting at Week 25 (the end week for the project) we can work backward by subtracting 2 weeks for Task 21 (bringing us to Week 23), followed by another 3 weeks for Task 20 (bringing us back to Week 20). At this point, we diverge off the critical path to Task 18. Its latest possible start time is Week 19 (that is $20 - 1$). Next comes Task 14. Its latest possible start time is Week 17 ($19 - 2$) and so on until you merge into the critical path.
5. *Calculate your slack times.* You determine slack time for each task by subtracting its earliest start time from its latest start time (see Figure 5.4). The difference is your slack or the amount of flexibility you have in terms of when you can start that activity. Remember you have no flexibility when it comes to your critical path tasks. There is always slack for noncritical path tasks, but be careful. Unforeseen events always occur and if you use up all your slack you are likely to find yourself under increasing pressure toward the end of your project.

Having calculated earliest and latest start times you will have clearer ideas of how speedups or delays will affect your project. This makes PERT/CPM a useful planning tool especially since project planning software programs helps expedite their

Task	Duration (Weeks)	Earliest Start Time	Latest Start Time	Slack
1. System Specification*	3	0	0	0
2. Define Data Dictionary*	3	3	3	0
3. Hardware Specification	3	3	7	4
4. Screen & Report Layouts*	5	6	6	0
5. Select Terminals	2	6	10	4
6. Select Computer	2	6	10	4
7. Design Test Procedure	3	6	14	8
8. Site Preparation	2	6	15	9
9. Computer Order & Delivery	5	8	12	4
10. Terminal Order & Delivery	5	8	12	4
11. Write Terminal Interface	3	8	16	8
12. Prepare Test Data	3	9	17	8
13. Select Database Manager*	3	11	16	5
14. Install Hardware	2	13	17	4
15. Develop Backup/Restore*	6	14	14	0
16. Program Screen and Reports	3	14	19	5
17. Program Database Definition	3	14	19	5
18. Test Terminals	1	11	19	8
19. Write Users Manual	2	17	21	4
20. Integrated Testing*	3	20	20	0
21. Train Users*	2	23	23	0
End	0	25	25	0

*Denotes critical path task.

FIGURE 5.4 Calculating earliest, latest, and slack times for designing and implementing a computer system.

creation. On the other hand, PERT/CPM is less useful as a control tool to monitor ongoing progress. For monitoring ongoing progress Gantt charts are more helpful because they enable you to see schedule variances more clearly.

5.1.3 Establishing and Maintaining Records

Your Gantt chart and/or PERT/CPM diagrams lay out your project development plan. Once it is established, you need to turn your attention toward keeping it up to date and maintaining other records that will help you monitor progress and learn from your efforts. The need for keeping records was presented in Section 3.2.2. In the next few paragraphs, additional factors and requirements are listed and applied to three engineering design projects.

Record keeping is usually a function of project management. To manage an engineering project competently, dozens if not thousands of records must be kept, analyzed, and managed. Here are just a few of them (with examples):

- Design criteria (weight, size, speed, etc), engineering standards (see Figure 2.1)
- Design calculations (strength required, weight)
- Design verification (doing modeling, scale model tests), computer programming (program documentation)
- Gantt, PERT, and CPM charts (tracking for project phases and timing)
- Document and component numbering (labeling wiring diagrams and labeling components on the device or system)
- Configuration management documents (tracking records, drawings, components, and people)
- Environmental protections and safeguards (Environmental Protection Agency air and water quality standards)
- Quality assurance requirements (manufacturing specifications)
- Licenses, permits and passes, requisitions and purchase orders, service contracts and repair orders (guarantees)
- Various engineering professional societies' design specifications and/or manufacturing codes that have been accepted as industry standards. Such organizations include American Society of Mechanical Engineers (ASME) III design specifications (piping and pressure vessel codes), Institute of Electrical and Electronics Engineers (IEEE), American Society for Testing and Materials (ASTM), American National Standards Institute (ANSI), and others.

If we consider three projects—the robot design project, the robotic arm project, and the design and construction of a fossil fuel power plant—we can see where these factors might be part of the design project and the record keeping process (Table 5.3).

However most record-keeping concerns, at least on the professional engineering level, revolve around program control or budgetary concerns. Table 5.4 gives a checklist developed by NASA that describes common causes for why engineering projects get out of control or exceed costs.

Note that, in each instance, good record keeping can help mitigate these problems. Good record keeping also comes in handy at the end of the project, at systems verification (fabrication, integration, test, and evaluation) or even during the operations phase when you may want to evaluate the entire project and report your results to the professor or client.

5.2 COMMUNICATION

In Chapter 2 we discussed basic composition skills as well as the importance of using correct grammar, spelling, and punctuation. We also discussed style, syntax, and diction and offered suggestions about how you can develop your own personal

TABLE 5.3 Records to Keep for Different Kinds of Design Projects

Design project records	Robot	Robotic arm	Power plant
Design criteria	X	X	X
Engineering standards and studies		X	X
Design calculations	X	X	X
Design verification	X	X	X
Computer programming	X	X	X
Gantt, PERT, and CPM charts	X	X	X
Document and component numbering	X	X	X
Configuration management documents			X
Environmental protections and safeguards			X
Quality assurance requirements		X	X
Licenses, permits and passes			X
Requisitions and purchase orders	X	X	X

writing style. Following, we present practical applications of those fundamental notions of good composition in some of the writing assignments expected of you in your engineering design courses. In particular, we examine proposal writing and ways to convey your design solution in a clear, concise manner.

5.2.1 Writing Proposals

Proposal writing can make or break a good design solution. The ultimate purpose of a proposal in the professional world is to sell a design solution to a prospective client. You may have formulated a good design solution, but if it is poorly expressed it can throw off the other design team members and cause your instructor to conclude that the project is poorly conceived or is not very well thought out.

TABLE 5.4 Reasons Engineering Projects Do Not Meet Time and Cost Goals

1. **Poor definition of requirements and objectives**: Have you and your team members clearly articulated what you must accomplish and what is required to accomplish it?
2. **Poor planning and review efforts**: Are you using a planning technique (e.g., Gantt) and are you systematically reviewing progress in relation to this plan?
3. **Inadequate WBS**: Have you clearly articulated the tasks and activities that need to be performed? Ideally try to break down tasks so that they are manageable and specific responsibility can be assigned and accomplished independent of other ongoing activities but at the same time can be integrated back into the total project and measurable in terms of progress.
4. **Starting/completing activities out of sequence**: Have you carefully prioritized your work and associated activities? Does the sequencing make sense and can you see how one step will lead to the next?
5. **Unforeseen technical problems**: Are you confident that you have done the background research necessary to know the technical challenges posed by your project? Do you know which problems people have encountered on similar projects?

There are two kinds of proposals: solicited and unsolicited. The former is a formal response to a request for proposal (RFP) or an invitation from the client. The unsolicited proposal emanates from the enterprising engineer who detects a client's need and seeks to address it with a design proposal. The analogy to the engineering design classroom is the difference between an assigned topic (an RFP) and an open topic, where students develop their own design projects. Either way, a good engineering design proposal always starts with a clear, concise statement of the problem, based on the real or perceived needs and desires of the client. Some professors also may ask for a cover page and/or an introduction and background before the problem statement is presented. A cover page typically includes your name, the project title, the name of the team, and the all-important date of the proposal. You also may be required to introduce or lead up to your problem by providing some background on how the design problem came about in the first place. Was there a specific need on the part of a client who sought your assistance? The rest of the design proposal includes material you can (or should be able to) take directly from your lab record notebook. Your professor might give you a list of topics to cover in a formal proposal, but Table 5.5 gives the more common ones used by professionals and engineering students alike.

We discussed items 1 to 5 from Table 5.5 in Chapter 2, but we will briefly touch upon them again. The introduction or background should essentially describe the context in which your design problem exists: What are the social, political, economic, environmental, and/or technical issues that are the basis for the design problem in the first place? The problem statement essentially summarizes this background information into a clear and concise statement(s) of the actual and/or underlying problem to be solved. Refer back to Figure 2.1 for an example of these two components and how they relate to one another.

The survey of the literature and the technical survey are discussed in Section 2.1.1 Recall that the survey of the literature is simply your research. Have you checked to see if anyone else has successfully completed a similar design project? What

TABLE 5.5 Contents of a Design Proposal

1. Introduction or background
2. Problem statement
3. Survey of the literature
4. Technical survey
5. Constraints
6. Alternative solutions
7. Statement of work
8. Expected results/contributions
9. Design schedule
10. Budget
11. Qualifications of team/members
12. References/bibliography

resources (articles, interviews, books, Web sites, etc.) are useful to you as you begin to propose your design solution? You can list those items in the references section or bibliography to demonstrate that you are well versed in the problem at hand.

The technical survey is similar to the literature survey, except you are researching the tools and techniques you will need to complete the design project. Your survey includes hardware, software, and standard procedures as well as proven engineering techniques needed to get the job done.

Constraints, as discussed in Section 2.1.3, can be many. The most obvious are the laws of nature, especially gravity, as well as friction, thermodynamics, and magnetism.

It is often said that design engineering is the reconciliation of a client's needs and wants with the realities of nature. Other obvious constraints are time and money, but you have to decide how much is enough. You can run into other kinds of constraints that are not so obvious, such as "conceptual blocks" and biases (see Section 2.1.2). One team of freshman engineers designed a perfectly functional solar-powered bicycle for disabled riders that would run on its own. However, in the eighth week of their project they realized that they forgot to build in enough power to carry the additional weight of a 120- to 200-pound person. We can only wonder if they then realized the concept of inclines and declines for power and braking.

As we stressed in Section 4.2.2., many of these constraints can be anticipated and dealt with by a shift of thinking. Vertical thinking, for example, begins with a theory and seeks a single solution. Lateral thinking generates alternative solutions by seeking a wider range of approaches. Alternative solutions are more likely to emerge from a shift in thinking than from repetitive efforts in forcing a solution. Recall the engineering team that developed several prototypes of tomato pickers before someone on the team took a look at the tomato instead of the machines. The result was a new tomato hybrid with a tougher skin and more accessible fruit. Digging for gold at the end of a rainbow may be enhanced by digging more holes than by digging them deeper.

This brings us to the next portion of a typical design proposal, alternative solutions. In this section, you will want to describe the various options that you considered, explain the criteria you used to evaluate them, and explain the rationale that ultimately has led you to recommend a particular solution or approach. As an example, take a look at excerpts from a design proposal for the autonomous robot project (Figure 5.5) The team starts with background and includes a problem statement as well. They then go into preliminary design concepts. In this part of their proposal, they allude to several alternative solutions for designing the chassis of their robot and provide a rationale for why they settle on an Erector Set and a PVC chassis. Under requirements and constraints, the student design team combines a technical survey with their list of constraints to convey the challenges they face in building a multifunctional robot.

You settle on one solution in the statement of work (SOW), the most feasible one. The SOW, also called method of solution or simply methodology, is the heart of a well-written proposal. The previous sections should lay the foundation for your SOW. They provide the rationale for why your methodology is the best approach. Then, in the SOW itself, you explain in detail what you will do and how you will do it.

(Part of a) ROBOT DESIGN PROPOSAL

I. Background

In the year 2020, much needed materials have been discovered on an asteroid? The materials are small orbs known as ultimatium and unobtainium. These materials must be collected, separated, and deposited into separate bins in order to be properly refined. There is one problem though—the terrain is inaccessible to humans. The only solution is to build a robot to do the job. This is where the engineers come into play.

It is the team's job to build a robot capable of maneuvering the terrain and mining the ore from the veins within a set amount of time and budget. The job will not be quite that simple, though, as other teams' robots will also be trying to collect the ore and may create more obstacles. After the robot is positioned, no interactions may take place between an individual and the robot until the robot exits the course. Some obstacles can be determined by IR signals being emitted, or by the color of the platform at the starting position.

II. Preliminary Design Concepts

The scenario presented is to be solved in the most efficient and creative way possible. There are also several factors to keep in mind while planning the construction of this robot. The budget is limited as well as the size of the vehicle. Secondly, there are a several small operations that make up the complete task. The operations include: starting and deciphering IR signals, positioning the vehicle and collecting marbles, sorting marbles, crossing the bridge, depositing the marbles, and exiting the course. In moving up inclines, staying on the correct path and/or correcting the path of motion, and avoiding the other robots.

In the chassis design, the team had to choose from PVC, cardboard and Erector set components. We recommend the Erector set because of its wide range of mobility, quantity of parts, and low cost. In addition, the Erector set parts can also come in handy during assembly of other robot components. We also propose that a piece of PVC be laid over the erector set frame after the motors are put into place to add even more stability. Although cardboard has a low cost it is very unreliable and, therefore, not considered for the chassis.

III. Requirements and Constraints

Budget: In the Robot Design Project, each group is allotted a budget of $140 to be used to buy parts for the robot. Since there would be bonus points awarded for every dollar not used under $140 dollars, a goal of $100 dollars was set. This would also allow $40 extra dollars in case an emergency situation ever arose.

Structure: All structural parts provide support and strength to the robot and the subframe. This includes all cosmetic structures.

Sensors: A mechanical or electronic device is used in gathering or distributing information within the environment.

Drive System: The direct energy transfer system goes from the power source (battery or fusion device) to the wheels (or application of force to the outside world). This includes all housing pieces, but does not include mounts to the chassis.

FIGURE 5.5 Proposal for the autonomous robot project. *(continued)*

Size/Shape: The robot, in its starting configuration, will be able to fit within a footprint of 9x11 inches and must be able to enter the door to the robot maintenance building which is assumed to be no more than 24 inches high. Any means may be used to propel the robot, providing that the robot is safe to operate and does not damage the course.

Starting Action: The robot will start at rest in the starting section of the competition course. It will be started by the recognition of a light signal emanating from below the robot.

Start Sensor: Each team will be issued a light sensor to be used as a start sensor. This light sensor should be located approximately below the robot's centroid. The controller program should use the light sensor input as the starting signal.

IR Emitter/Sensor: The course will be outfitted with one or more IR beacons that may prove useful in navigation. Each team will be supplied with one IR receiver and may purchase additional IR receivers capable of detecting the beacon(s). Under no circumstance may a robot emit an IR signal.

Robot-to-Robot Interaction: Robots may be designed to interact with the rules set down in the section governing the robot-to-robot interactions during the head-to-head competition.

Loose or Disposable Parts: Any parts that are intentionally or unintentionally dropped or lost during a single run will be confiscated for the rest of the contest. (Our team will need seven copies of any part intentionally dropped, and the cost of the seven copies must be included in our robot cost.)

FIGURE 5.5 *(continued)*

The design schedule is often combined with the SOW so we get this formula: who does what, and when is it done? Keep in mind three basic components of a project—performance, budget, and schedule. The engineering design student can clearly see all three elements running through the proposal. The successful design project is on time, is at or under budget, and, above all, it works!

The design schedule, incorporated into the SOW or standing separately, shows clearly each task to be completed, the team member responsible, and the date of completion. To get to that point, team members brainstorm and list all the activities to be accomplished over the lifetime of the project. Then the sequence of the activities is set, along with the amount of time needed for each task, often in a Gantt chart. The final step in this process of building a design schedule is to agree who will be responsible for each and every task. When this process takes place with the entire team together, each member has "buy-in," ownership of the design schedule, and commitment to it. Does that mean the best-qualified person does each job? Not necessarily. In professional training and development, engineers are often given job rotations to experience new challenges and expand their range of capabilities. The most important criterion is a genuine desire or determination to complete the task and to do it well.

Of course, each schedule should have some slack built into it, just as there should be some reserve built into the budget for unexpected events. The schedule also should be marked for major events and milestones. This effort will pay off greatly when you are asked to do progress reports, showing any slips or jumps in schedule. A slip in schedule usually results in an increase in budget. Professionally, it could also result in project cancelation or fewer new business opportunities later. That's why professors are so firm about deadlines.

For professional engineering projects, budget preparation is often a complex process that includes consideration of many factors: time, salaries, risks, material, overhead, inflation rates, and transportation, to name just a few. In terms of most student design projects, the following are some typical budgetary items you may want to include in your proposal: material costs, lab fees, research expenses, and communication charges (e.g., fees for phones and Internet access). Although your budget is not complex (compared with many professional engineering projects), you still should try to include one in your report. See the sample robot project budget in Figure 5.6.

Under expected results you summarize the benefits of your doing the work. This might include a design that is superior to others or the solution to a problem that is of critical interest to the client (or instructor). If the proposal were being written to industry, it would summarize the profit to be made, costs to be saved, and perhaps the increased user interest (market share) because you are doing the proposed work.

The qualifications of team and members explains why you are the best people to take on your particular project. Of course, as students you may feel hard-pressed to articulate your unique qualifications. After all, engineering design is new to you. Take a stab at writing this section anyway. It is good practice and you also may be surprised with what you come up with. For instance, perhaps you have had relevant experience with aspects of the design problem you are facing. Maybe you have a particular expertise in math and science. Perhaps you can cite relevant qualifications based on your hobbies or even through exposure you have had from family and friends. Do not be afraid to "toot your own horn." Your design proposal is a good opportunity to practice your ability to convince others of your expertise.

One final comment about writing proposals must include the value and importance of graphics. Simple charts, properly labeled, can show data and information in ways that mere words and sentences cannot. Diagrams are useful in proposals, especially wiring diagrams, and pictures are worth a thousand words each. Software programs such as Quattro Pro and Adobe Illustrator can enhance the written words of a design proposal greatly, but first-year students often cannot afford the time to master them. Upper division and graduate students should have them in their toolbox of design tools and techniques. And don't forget the Gantt chart. At a glance your professor or client can see a logical, methodical plan of attack.

5.2.2 Preparing Bibliographies

The last piece of your proposal is references or bibliography. Most professors and some clients may wish to see where you got some of the information contained in your proposal. Following is a brief guide for preparing this portion of your design proposal.

First, you can ask your client or professor if any particular bibliographical style is preferred. Some professors may want you to learn and practice the commonly accepted documentation style for civil, electrical, mechanical, and chemical

Beginning Budget Amount—$140 Projected Final Remaining amount—$30.00			INCOME		
Purchases as of Date (Final)					
Date	Product	Quantity	Cost	Total Cost	Balance
4/12/99	PVC—55.95 sq inches	1	$1.80	$0.70	$139.30
4/12/99	Twin Motor	1	$14.95	$14.95	$124.35
4/12/99	Tank Treads	2	$4.29	$8.58	$115.77
4/12/99	Wheel Assembly Main/Rear	1	$3.69	$3.69	$112.08
4/12/99	Wheel Assembly Main/Front	2	$4.89	$9.78	$102.30
4/14/99	Diodes	28	$0.00	$0.00	$102.30
4/14/99	PVC—24.8 sq inches	1	$1.80	$0.31	$101.99
4/15/99	Wheel Assembly Main/Rear	1	$3.69	$3.69	$98.30
4/16/99	PVC 5.02 sq inches	2	$1.80	$0.13	$98.17
4/16/99	PVC—4 sq inches	2	$1.80	$0.10	$98.07
4/16/99	Bush Wheel	2	$2.25	$4.50	$93.57
4/16/99	Shaft Encoder	1	$1.65	$1.65	$91.92
4/16/99	PVC—8.82 sq inches	1	$1.80	$0.11	$91.81
4/20/99	Eye Bolt	2	$0.29	$0.58	$91.23
4/20/99	Nuts/Screws	2	$0.10	$0.20	$91.03
4/21/99	Right Angle Erector Part	2	$0.29	$0.58	$90.45
4/21/99	7-Hole Perforated Strip	2	$0.56	$1.12	$89.33
4/21/99	1 ½ 3-Hole Erector Part	4	$.030	$1.20	$88.13
4/21/99	Nuts/Screws	1	$0.10	$0.10	$88.03
4/21/99	Male Header	1	$1.29	$1.29	$86.74
4/21/99	Nuts/Screws	8	$0.03	$0.20	$86.54
4/22/99	Couplers	2	$3.24	$6.48	$80.06
4/27/99	Female Header Strip	1	$1.29	$0.07	$79.99
4/27/99	Erector 2-Hole	2	$0.33	$0.66	$79.33
4/27/99	Nuts/Screws/Washer	6	$0.03	$0.15	$79.18
4/30/99	Flex Plate—2.5 x 2.5	1	$1.27	$1.27	$77.91
5/2/99	Microswitch	3	$1.76	$5.28	$72.63
5/2/99	Mercury Switch	1	$0.75	$0.75	$71.88
5/2/99	Servo Motor	1	$10.00	$10.00	$61.88
5/2/99	PVC—43.16 sq inches	1	$1.80	$0.54	$61.34

FIGURE 5.6 Sample budget from robot design project. *(continued)*

5/4/99	Male Header	1	$1.29	$1.29	$60.05
5/10/99	PVC—29.59 sq inches	1	$1.80	$0.37	$59.68
5/11/99	PVC—2.40 sq inches	1	$1.80	$0.03	$59.65
5/12/99	Dow Rod	1	$0.32	$0.16	$59.49
5/13/99	Transparent Plate 2.5 × 2.5	1	$1.53	$1.53	$57.96
5/13/99	Washer	1	$0.03	$0.03	$57.93
5/13/99	Cardboard	1	$0.10	$0.10	$57.83
5/14/99	Servo Motor	1	$10.00	$10.00	$47.83
5/14/99	PVC - 20 sq inches	2	$1.80	$0.50	$47.33
5/17/99	Microswitch	2	$1.76	$3.52	$43.81
5/20/99	Magnets	1	$0.25	$0.25	$43.56
5/25/99	Magnets	2	$0.25	$0.50	$43.06
5/25/99	Nuts/Screws	20	$0.30	$6.00	$37.06
5/25/99	Velcro	1	$1.00	$0.10	$36.96
Provided Materials—Duct Tape, Electrical Tape, Soldering Iron/Solder, Welding Iron/PVC Solder, Scotch Tape/Labels, Glue Gun/Hot Glue, Shrink Tubing, Drive Train Axle Sleeves, Diode Board/Diodes/Rectifiers, Motor Chips/Heat Sink, Paperclips					

FIGURE 5.6 *(continued)*

engineering, depending on your project and area of concentration. There are some standard style guides, however, that you may have used in high school and general distribution courses in college.

Internationally, the most common documentation method for engineering journals is the style of the Institute of Electrical and Electronic Engineers (IEEE). The in-text citations use brackets that surround a number referring to a reference at the end of the text [42]. Only the essentials are used in footnotes and references are listed numerically at the end with commas instead of periods:

- 42. Author, U.R. *Title of Book*, City: Publisher, Year.
- 43. Author, U.R. "Title of Article," Journal vol.#, no.#, pp.#-#, month, year.

The Modern Language Association (MLA) style is another approach that is commonly accepted. With MLA style, in-text citations are streamlined (Ferguson 42) and truncated when the author is mentioned in the text (42). In the references section at the end, authors are listed alphabetically like this:

- Ferguson, Eugene S. *Engineering and the Mind's Eye*. Cambridge: MIT Press, 1992.

A third common alternative is the style of the American Psychological Association (APA). For this style, citations are similar (Ferguson, 1992, p. 42) but references are placed in a works cited page at the end in a different style:

- Ferguson, E.S. (1992). *Engineering and the mind's eye*. Cambridge: MIT Press.

Regardless of the style used, consistency and user friendliness are key to preparing bibliographies. Unless prescribed otherwise, find a style you are comfortable with and be consistent. Use the IEEE, MLA or APA stylebooks, and make sure the professor or any reader can access the same information you cited or used. After all, that is the main function of a bibliography. Think of it as a guide for further reading in a specific area.

5.3 COLLABORATION

There are two important and related outcomes from effective collaboration during the formulating solutions stage. The first is an open and inquisitive climate within your team. Without an open climate people may feel inhibited, frustrated, or even distrustful. When these feelings dominate a team's dynamics people withdraw and/or focus on their individual needs instead of on what is best for the team. Contributions fall off, efforts diminish, and communication breaks down. As a result, a team fails to fully explore all possible solutions and may move into design Phase 3, developing models and prototypes, with a poorly conceived plan.

Second, effective collaboration increases the likelihood that all team members will feel committed to the solutions you identify. Remember, even when you complete this phase of the design process you are still at a relatively early point in your design efforts. Much remains to be done before your design is transformed from an abstract idea into a viable product, service, or process. Without enthusiastic commitment from all team members, it will become increasingly difficult for ideas to be implemented and developed further. Therefore, all team members need to feel comfortable with the decisions reached and all should feel they played an active part in making them.

We begin this section by providing guidelines on what you can do to encourage full participation and to build commitment within your team. We will next turn your attention toward some related techniques for handling two challenges all teams face: managing conflict and avoiding groupthink. Understanding and applying these concepts will help ensure that your team makes the most of everyone's participation now and in the future.

5.3.1 Ensuring Full and Open Participation

Recall that some of the great benefits of working on teams include more ideas and different sources of knowledge. However, these sources of knowledge are useful only if they are tapped. In fact, encouraging everyone to participate may be the most critical piece of team collaboration; if ideas go unheard, they cannot be evaluated or used.

Ensuring full participation is the responsibility of everyone on the team. Here we have listed some suggestions about how the group can work together to ensure that everyone participates. Additionally, there are steps individuals can take to be sure they are encouraging participation at every level.

Some Guidelines for Ensuring Open Participation There are two ways to think about ensuring open participation within your team. On one level it helps to think about what your entire team should be doing as a unit. At the same time you also need to think about the things you as an individual should be doing.

What can the team do? The best way for your team to promote open participation is by establishing ground rules and procedures within your working agreement (see Section 2.2.2.3) that promote information sharing and involvement. Here are some examples:

- On a rotating basis, appoint different team members to run team meetings.
- Make sure you have an agenda for team meetings, ensuring that everyone agrees on the topics to be discussed. This doesn't have to be anything fancy; just a handwritten list of topics is sufficient.
- Create and review checklists of topics that need to be covered and criteria that need to be met.
- Specifically ask for participation from any team member who has not contributed during a previous meeting.
- Establish time during meetings to summarize what each person has said and ensure that feedback is given on each suggestion.
- Establish specific work roles whenever possible, and make certain people know what is expected of them from one meeting to the next.

What can individuals do? One of the causes of weak participation is the tendency for one person to dominate the whole conversation. People who dominate meetings often do not realize they are doing it, especially when they are excited about an interesting project or discussions. To assess your tendency to dominate discussions within your design team, read the list in Table 5.6 and determine whether or not it describes your behavior. If it does, consider using the suggested tips to reduce your tendency to dominate, and encourage others to open up and participate. If each member of the team takes steps to listen to all other members and encourage participation, collaboration among team members will improve dramatically.

There are other things all team members should do regardless of whether they have a tendency to dominate conversations. For instance, realize that you have an obligation to participate and that others are expecting your input and efforts. If you find yourself being reticent or nonparticipative, try to determine why. Are you concerned about having your contributions rejected? Perhaps you think you do not understand the topics well enough. Maybe you are just feeling overwhelmed with too many things on your plate, or perhaps you are just having a bad day. Once you understand why you are not participating, then you will be in a better position to do something to change the situation. This might include explaining your situation to team members and asking for their help. Even if they cannot help, addressing the issue with others will reinforce the fact that full participation is an important team value.

In his book *The Skilled Facilitator*, Roger Schwarz describes two other team behaviors that help promote participation:

- Make statements and then invite questions and comments. This means expressing your views but then inviting others to respond, whether they agree or

TABLE 5.6 Examples of Dominating Behavior and What to Do about It

You know you tend to dominate if you:
- Interrupt frequently to interject your opinions.
- Restate your opinion often.
- State your opinion forcefully.
- Speak often, preventing others from having a chance to participate.
- Criticize the ideas of others too quickly.

Steps you can take to reduce your tendency to dominate:
- Resist the urge to interrupt.
- Limit the number of times you will state the same opinion unless specifically asked to repeat it.
- Start your statements off with "I think," not "Everyone knows" or "It's so obvious."
- Be aware of the proportion of the time you spend speaking compared with others in the group; if you are using more time than others, give them a chance.
- Refrain from hasty judgment.
- Ask open-ended questions that will encourage others to participate.

Encourage others to express contrary opinions by:
- Waiting until others finish speaking, even if you are sure you understand the argument.
- Restating the main points of the other person's point of view to ensure you fully understand the perspective.
- Asking the person to verify the accuracy of your restatement and clarify it if necessary.
- Identifying the points about which you agree.
- Stating the points you disagree with and why, if appropriate at the time.
- Not engaging in side conversations while someone else is giving an opinion.
- Focusing your attention on the speaker.

disagree. For example, a team member might say "I think it might help to build a beer tap into our dorm room design. But some of you might feel differently and I would like to hear what each of you thinks about my idea, even if you disagree." Inviting people to comment on your statements encourages discussion that is a dialogue rather than a series of monologues.

- Do not take cheap shots even if you think they are witty. Almost all of us have been the target of a snide remark at some point. Aside from the fact that the recipient of such remarks is made to feel bad, there is another practical reason for avoiding them. The insulted person usually winds up being distracted. Instead of focusing on the work to be done, the victim of insults is more likely to spend time feeling hurt or angry, wondering why the comment was made, or thinking up a clever comeback. The bottom line is that once someone is distracted, that team member cannot participate.

Finally, even when not intending to insult someone we can all say things that discourage others from participating. When responding to others, try to avoid using "conversation killers" like the ones included in Table 5.7.

TABLE 5.7 Conversation Killers

Words and phrases	Gestures and actions
• That's a dumb idea. • Don't be stupid. • They won't let you. • I have a better idea. • It will never work. • You just don't get it. • Yes, but...	• Deathly silence • Quickly changing the subject • Quickly offering another suggestion • Shaking head in disagreement while someone is speaking

5.3.2 Reaching Consensus and Building Commitment

During the formulating solutions stage, full participation is critical, but remember it is a means to an end and not the end result itself. Your ultimate goal during this stage is to identify specific alternatives and solutions for your design problem. Earlier, Section 4.3 focused on how your team can decide which alternatives are best. This section focuses on the impact of decision making on team dynamics, specifically making sure everyone agrees to and supports the best possible solution.

Structuring a Fair Decision-Making Process One indispensable way of getting buy-in and commitment from team members is to ensure that the process is fair to all participants. If team members believe the process is unfair, then they are more likely to reject the decision and will fail to support it. Acceptance and support are critical, especially when the time comes for team members to contribute more resources to creating models and prototypes and to do work necessary for presenting a completed design project. Energy invested early during the formulating solutions stage is well worth the payoff in the end. Try to follow the guidelines described in Table 5.8 in order ensure that your team's decision-making process is fair.

TABLE 5.8 Ensuring a Fair Decision-Making Process

What	How
Opportunity for participation	All team members have an equal chance to participate.
Consistent rules followed	All team members agree on basic criteria; all ideas are evaluated against criteria.
Bias not a factor	All team members have an equal chance to participate in both idea generation and evaluation of ideas.
Correct information used	Alternatives are fully understood by all team members.
Opportunity to reconsider	Team will reconsider a decision when new evidence arises that a different alternative is more viable.

Using a systematic decision-making grid (such as the example discussed in Section 3.1.3) can be a helpful tool for ensuring that your team's decision process is fair and objective. The grid forces you and your team members to evaluate alternatives in relation to specific and objective criteria. As a result, it helps keep discussions focused and minimizes the impact of personal biases.

Reaching Consensus An objective and unbiased decision process is also more likely to result in a consensus-based decision. Consensus is vital for any decision that must be supported by all team members. This is exactly the kind of support required by student design teams trying to identify design solutions. But what does consensus really mean? If you think consensus means 100 percent agreement, then it is likely to be a concept that confounds rather than helps your efforts to collaborate with other team members. A more realistic definition of consensus focuses on the process by which a decision is made. In other words, your team can say it has reached consensus when all team members agree that 1) "I have had an opportunity to express my views and they have been fully considered by the team" or 2) "Our solution may not be ideal, but I think it will work and I can support it." This means all team members have an obligation to express their opinions as well as to listen to the opinions and feelings of others.

5.3.3 Managing Conflict

As you explore design solutions and work toward consensus, each member of your team will have opinions, expertise, and ideas about aspects of the design problem. Each alternative will need to be considered, trade-offs weighed, and competing points of view worked through. Conflict, of course, can be difficult, leaving participants feeling frustrated and full of stress; however, conflict that is managed well can be rewarding because it helps bring out various perspectives and can lead to solutions that are better than any individual could devise alone.

Managing conflict requires that individuals understand the benefits of working as a team while formulating solutions. As noted already, one benefit of working as a team is the potential for the solution to be better than any solution an individual could come up with alone. There are several reasons why a team can develop a better solution than individual participants. A team has the benefit of more participants with broader and more diverse knowledge to contribute to the solution. Additionally, more participants have the ability to understand and make sense of information that will be useful in making decisions. Of course, these benefits arise only when the expected conflict is managed well by those in the group.

As important as it is to understand the benefits of working on a team to formulate solutions, it is equally important to understand the potential pitfalls of teamwork. If conflict is not managed well, then some members on the team may not fully share information and other resources, with the result being less-than-optimal solutions. Techniques for managing conflicts are given in Table 5.9.

The first step in managing conflict is understanding that there are different approaches to dealing with conflict. The best possible outcome is that members engaged

TABLE 5.9 Managing Conflict

What	How
Acknowledge that conflict is normal and is to be expected.	Build in time for debate; ask for feedback and debate.
Acknowledge personal biases.	Use active listening; judge content, not delivery; use others as a sounding board.
Understand project definition and constraints.	Review tasks to be done; assess resources needed to complete tasks; determine who does what and when.
Encourage open participation.	Ask others to contribute; listen to all alternatives.
Acknowledge that participants are new to the task.	Take time to know task; take time to get to know strengths and weaknesses of participants.
Seek to understand ideas of others.	Before getting those around you to understand your perspective, try to understand theirs. You might realize you are closer to theirs than you thought.
Don't always be right.	Acknowledge that others have good ideas. Don't immediately correct others; tolerate imperfection.

in the conflict think they have found a solution that works for both the group and their individual requirements. Arriving at such a win-win solution requires that everyone participating deal with the conflict in a constructive manner. Table 5.10 gives a quiz to help identify your conflict management style.

If you answered yes to any of questions 1 to 4 in Table 5.10, your team may be avoiding conflicts.

Teams that avoid conflict may digress in their conversations. Take notice if team members bring up some other topic (e.g., what they did last night, the most recent calculus test) in an effort to divert attention from an issue that might be difficult to work through. Digression from a difficult topic may result in the same issues coming up over and over again and being ignored each time. Ask yourself as you leave team meetings if you know what you should be doing in preparation for the next meeting. If you are not clear, seek out other team members to determine if they know. Confusion on the part of all or most team members is a clear way to spot avoidance behaviors. Table 5.11 suggests steps to take if your team is avoiding conflicts.

If you answered yes to any of questions 5 to 8 in Table 5.10, your team may be too quick to accommodate.

Teams that are accommodating are not much different from those that are avoiding. In fact, accommodating is really a form of avoiding. The difference is that, in accommodating, the team members make efforts that may look like they are managing conflict. For instance, team members may believe that they are managing conflict by allowing their own opinions to be subordinated to another idea before both ideas have been fully explored as possibilities. Subordinating some ideas allows others to dominate the decision making—a sign that others are accommodating. It can appear that a team is managing conflict if they are allowing ideas to be presented—but they may be accommodating if those ideas are not explored. Table 5.12 suggests steps to take if your team is too accommodating.

TABLE 5.10 Quiz to Assess the Conflict Management Style of Your Team

Answer **yes** or **no** to each of the following questions.

1. Do you or your team members accept solutions and ideas without thoroughly discussing the pros and cons?
2. Do you leave team meetings without fully understanding what is to be done next or why?
3. Does your team keep having to deal with the same problems?
4. Do team meetings stay focused on the task at hand?
5. Do you or your teammates present a position and then immediately back down?
6. Do you sense that you or others feel uncomfortable saying what you really think or feel?
7. Do you or a few others on the team tend to dominate discussions and planning?
8. Do you or your team members believe that keeping everyone happy is more important than finding the best solution?
9. Do you or your team members blame others when things do not go as planned?
10. Are you part of a clique or subgroup that sticks together, regardless of the issue?
11. Do you or your team members show reluctance in considering alternatives that you did not contribute?
12. Do you or your teammates interrupt, or talk over, others?
13. Do you or your teammates lecture others in order to convince them you are correct?
14. Does your team decide major issues by voting?
15. Does your team attempt to satisfy everyone by incorporating everyone's suggestions, even if they weaken the solution?
16. Do you believe that most of the decisions your team makes are less than ideal?

TABLE 5.11 Some Steps to Take If Your team Is Avoiding Conflicts

- Decide to explore fully the pros and cons of each issue that comes up, even if it means an extra long meeting or some extra stress on the group for awhile.
- Before the meeting closes, have team members summarize the next steps for which they are responsible as well as the rationale for those steps.
- Start meetings on time.
- Create an agenda of the issues to be covered for that meeting.
- Appoint a team member to be responsible for keeping the discussion on track.

TABLE 5.12 Some Steps to Take If Your Team Is Too Accommodating

- Have team members who present an argument for or against an issue fully explain their rationale and defend it against counterarguments.
- Start discussions about the pros and cons of an issue by having team members write down their support or argument on paper, and then have each contribution read aloud and discussed.
- Appoint a different person for each meeting to act as the facilitator; this person should ask for each person's inputs.
- Evaluate each idea against the criteria for a good decision and not just because it was suggested.

TABLE 5.13 Some Steps to Take if Your Team Is Fighting

- Refrain from passing judgment or assessing blame if things do not go right.
- Divide cliques and meet in different subgroups to discuss issues, so everyone sees different sides.
- Ask members to defend an idea they disagree with in an effort to get them to see the positive sides of an alternative they did not consider.
- Spend one meeting reviewing the principles of active listening.
- Remind team members that each idea needs to be evaluated against the criteria of the best solution and not genesis of the idea.

If you answered yes to any of questions 9 to 13 in Table 5.10, your team may be in a fighting mode.

Fighting is sometimes signified by subgroups or cliques existing within the larger group. Sometimes these cliques start because of friendship or because of an initial agreement on a particular issue. In any case, if these cliques begin opposing ideas contrary to their own simply because they are contrary, then the group is fighting. A group that is managing conflict well may find situations in which a subgroup becomes very vocal about one particular point or about some issue, but when the same subgroup bands together repeatedly regardless of the issue, it may be an indication that the group is fighting. The style in which group members choose to present their ideas also may be indicative of whether they are fighting. If team members tend to lecture others, stating their opinions without opportunity for questions and discussion, then they may be fighting. Similarly, if certain team members tend to interrupt and speak over others, then they certainly are not engaged in active listening. Table 5.13 gives steps to use if your team is fighting.

If you answered yes to any of questions 14 to 16 in Table 5.10, your team may be too quick to compromise.

Like accommodating, being too quick to compromise is a method of avoiding conflict. Again, compromise can look like a good way to manage conflict, and in fact it becomes a problem only when the team is too quick to follow up on issues to the fullest extent. Teams that immediately settle disputes by voting may be missing valuable contributions that minority members may have. In a similar method of avoiding conflict, teams will add design features that don't really belong simply to satisfy a few members of the team. It may appear that the group is managing conflict, but if design solutions often fall short of expectations, they may be too quick to compromise. Table 5.14 gives suggestions for when your team is too quick to compromise.

TABLE 5.14 Some Steps to Take If Your Team Is Too Quick to Compromise

- Refrain from taking a vote to decide an issue, even if it increases the discussion time.
- Appoint a team member to the role of results checker—after a decision is made, this member walks the team through the evaluation criteria and compares the decision to these criteria.
- Encourage debate by having each team member state the pros and cons of one alternative, and have the rest of the team respond.

TABLE 5.15 Signs That Your Team Is Collaborating

- Team members feel free to communicate openly with one another.
- Team members listen actively.
- The criteria for a good solution are what drive the discussion, not frustration or anger and blame.
- Everyone understands that the best solution is best for the group regardless of where the ideas came from.
- All alternatives are explored, and alternatives are combined to create even better solutions.
- Everyone understands the steps in the process and agrees about what is the best next step.

Perhaps you took the quiz and answered "yes" to hardly any of the questions. In that case, your team may simply be managing conflict in a productive manner. Collaborating is a significant accomplishment, especially in teams that are new to each other and new to the tasks they need to complete. Read through the list in Table 5.15 and determine whether your team is collaborating. If you don't see evidence of these behaviors, then go back and reevaluate the answers you gave to the quiz, being as honest as you can. If you do see these behaviors, then congratulations on a job well done.

5.3.4 Avoiding Groupthink

We have just finished discussing how conflict can be productive, how to manage it, and how to ensure that everyone gets a chance to participate. Effective conflict management can also help prevent a different, but related, group phenomenon called groupthink. Imagine a team in which people are regularly yelling at one another and are in constant disagreement. If this behavior persisted throughout the team's life, we might predict that they are not going to accomplish very much. Now imagine a group in which there is seldom, if ever, any open disagreement. Although all team members possess relative expertise, they consistently follow the plans of action as laid out by one or two influential members. This team is also likely to be headed for failure and may be experiencing groupthink.

Groupthink is the kind of thinking that occurs when a team spends too much energy maintaining cohesiveness and solidarity instead of openly evaluating facts. Groupthink can result in very bad decisions. In most cases, these poor outcomes have very little to do with the intellectual capabilities of the team. Instead they are the results of subtle social and interpersonal pressures that discourage team members from making the best use of their talents. Certain symptoms and warning signs can indicate that a team may be the victim of groupthink:

- Members are highly cohesive to the point where being a member of the group is more important than being open about difference of opinion. No one wants to disagree for fear of being ostracized.
- The team is isolated from external information—members do not often seek or hear alternative viewpoints.

- A leader or junta controls the discussion very tightly and censures dissenting views through open rebuke or subtle pressure.
- The team perceives an outside threat and/or pressures to its existence or success.
- No standard method is in place to consider alternative viewpoints.
- The team has so many successes that they begin to feel cocky, as if they can do no wrong.

In his research paper *Red Flags, Smart People, Flawed Decisions: Morton Thiokol and the NASA Space Shuttle Challenger Disaster*, Dr. Mark Maier provides a well-documented study on managerial processes and the potential disastrous effects of groupthink. He writes:

> *On January 28, 1986, the Space Shuttle Challenger lifted off on the 25th mission of NASA's shuttle program. Its voyage was tragically cut short by an explosion just 73 seconds into flight. All 7 crewmembers perished.*
>
> *In its final report, issued June 10, 1986, the (Presidential) Commission noted that the immediate, technical reason for the disaster was the failure of O-rings in the right Solid Rocket Booster field joint to seal properly due to the freezing temperatures preceding lift-off. The Commission singled out the "flawed decision-making process" which led to the launch as the "contributing cause" of the accident.*

The Space Shuttle Challenger was scheduled for liftoff on January 28, 1986. The projected temperature for launch time was 29°F with overnight lows forecast

(a)

(b)

FIGURE 5.7 (a) The Space Shuttle Challenger at launch pad. (b) Moments later, after the explosion, groupthink may have contributed to this disaster. (Courtesy NASA.)

in the teens. Rocket engineers for Morton Thiokol had already determined that the O-rings that seal the joints as the solid rocket booster pressurize and ignite for launch exhibited fatal flaws. In particular, it was determined that at low temperatures the O-rings exhibited significant variation in performance characteristics.

While Thiokol was well aware of the dangers associated with the O-ring erosion, they still moved ahead with the flight schedule. Why? In this section, we are going to discuss some of the factors related to groupthink that directly contributed to the fatal decision to launch, resulting in the explosion and death of all crew members. We recognize that the decisions your design team are about to make may not have the tragic implications of the Challenger, but we are going to ask you to reflect on the dynamics within your own design team to determine whether any of the symptoms of groupthink are present.

Outside Pressures As noted earlier, a team perceiving an outside threat to its existence or success is highly susceptible to groupthink. This pressure can result in decisions being made that are less than optimal. In Thiokol's case, Congress was pressuring NASA to seek bids for an additional source to supply shuttle boosters. Thiokol's exclusive contract worth over one billion dollars was at risk. It was in Thiokol's best interest to move forward with the launch schedule.

If there are potential outside pressures on your project, you should answer the following questions:

1. Are there any outside pressures your design team is facing that might contribute to groupthink (e.g., time demands, grades, other courses, or social life)?
2. Can you think of any ways to help minimize the impact of these outside pressures?

Too Much Cohesion/Fear of Being Excluded As we discussed in Section 2.4 during the accepting stage of team development, conflict is giving way to cohesion and camaraderie. We most often think of cohesion in positive terms. However, there is also a downside to cohesion. Members of a team may feel intimidation or fear of alienation if they do not conform to the team's norms. For instance, engineers responsible for the O-ring design from both Thiokol and Marshall Space Flight Center initially voiced objections to management about the pending launch. Although they knew their evidence was not 100 percent conclusive, they felt strongly that it presented enough of a trend to warrant postponing the launch, at least until the weather became warmer. Initially, they were quite adamant but eventually backed down in the face of strong censure from their managers. Here is a brief description of what happened:

> *Engineers Arnie Thompson and Roger Boisjoly [the engineers who designed the O-ring seal system], anxious that their recommendation might be reversed, argued with the managers and urged them to stand by their original decision [not to launch]. Thompson was the first to attempt this. He got up from his position at the end of the table and sketched the problems with the joint on a note pad in front of the managers. . . . But the senior rocket engineer abandoned his effort after Mason [Thompson and Boisjoly's manager] gave him an unfriendly look as if to say . . . "Go away and don't bother us with the facts."*

TABLE 5.16 Team Cohesion/Fear of Being Excluded

1. Are people on your design team reluctant to contribute ideas because they don't want to go against the group or "rock the boat?"
2. If some members voice a contrary opinion, does the rest of the team put pressure on them to change their minds? If so how does it occur?
3. Does your team openly evaluate contrary or minority opinions?
4. What, if anything, does your team do to encourage constructive debate and discussion?
5. What might your team do to ensure more open participation among all team members?

> *Boisjoly then stood up from his chair across from the managers, slammed the photographic evidence from previous flights down in front of them and literally screaming at them, admonished them to "look at what the data are telling us: namely that cold temperature increases the risk of hot gas blowing by the joint." He stopped abruptly when Mason glared at him icily with "the kind of look you get just prior to being fired."*

When later queried about why they did not press further, both engineers cited respect for managerial authority and management's right to make the final decision. As the above excerpt suggests, the two engineers believed that if they protested further they might be fired from the team altogether. What might the result have been if the engineers had gone outside the chain of command to continue their objections to the launch, or if their manager did not attempt to silence them through the subtle but effective pressure he applied? Table 5.16 provides questions to ask to assess the extent that team cohesion/fear of being excluded dominate decision making.

Leadership Style This can be another factor contributing to groupthink. Leaders can control the decision-making process in a number of ways. They may state personal opinion before asking for input from others. Or, they may raise only certain questions or ask for input from only certain participants. This appeared to be the case with Challenger. At one point during the discussions, Jerry Mason, Senior Vice President at Thiokol, asked his managers, "Am I the only one who wants to fly?" Then, at a crucial point in the decision-making process, he turned to the sole holdout among his four senior managers and challenged him to "take off your engineering hat and put on your management hat." The engineers in attendance at the meeting were not included in the final "launch/no launch" vote. The four senior managers agreed unanimously to launch. Table 5.17 summarizes questions to determine whether leaders are contributing to groupthink.

TABLE 5.17 Leadership

1. How is leadership shared within your team?
2. Has someone on your team assumed the leadership role? Is the leadership style controlling?
3. Is there anything your team can do to ensure that team members are not unduly influenced by subtle pressures ("Am I the only one.... ?")?

TABLE 5.18 Guidelines to Avoid Groupthink

- Break into subgroups that meet separately to discuss different ideas openly.
- Reconvene to address viable alternatives.
- Go to an outside, impartial source who can provide fresh insight.
- Revisit the process used to evaluate the alternatives to see if it has bias.
- Listen to external suggestions and evaluate them against your standard criteria.
- Encourage ideas that are completely contrary to those accepted by the team as a whole.
- Identify the consequences of choosing an alternative route and determine just how costly they might be.

Earlier in this chapter we discussed techniques that will help eliminate some of these symptoms of groupthink. For example, getting all ideas heard will encourage alternative viewpoints, evaluating ideas against established criteria will help good decision making, and efforts to rotate members of the team will help reduce the possibility of a directive leader emerging. Nonetheless, groupthink is still possible. Is your team working so hard to avoid conflict that you are eliminating the constructive give-and-take that often leads to highly successful outcomes? Table 5.18 gives guidelines for avoiding groupthink.

CHAPTER REVIEW

As you identify and analyze various design alternatives, expect conflict and different points of view among your team members. It is essential for you to work through these differences constructively because you still have much work ahead of you. Recognize that the solutions you derive will also dictate the bulk of the work you will do next. Therefore, this phase is also the point at which you establish a detailed plan of action. Consider establishing a Gantt chart and using it to mark progress. If your project is more complex a PERT diagram may help.

In terms of collaboration your team probably is still at a relatively early stage in its development. Just an awareness of potential pitfalls such as groupthink can help ensure your team does not fall into that type of counterproductive pattern. Remember that better-quality ideas and solutions are more likely to occur when you ensure open and full participation of all team members. Expect conflict to occur and welcome it as an opportunity to develop ideas and to see things from different perspectives.

At this point your professor may ask you to write a formal proposal for your design project. Focus on making that proposal a clear and engaging document. It should demonstrate your in-depth understanding of the design problem, convey the rationale for your solution, and convince the reader that you have a clear plan of action for getting the work done.

To help you review this chapter, answer the following questions. In addition, review the behavioral skills checklist at the end of this chapter and use it to assess how well you are currently applying the skills and tools discussed in this chapter.

REVIEW QUESTIONS

1. The table below is a work breakdown structure (WBS) for designing the chassis for an autonomous robot. A) The tasks are not listed in the correct order. Establish a correct order for completing them. B) Prepare a Gantt chart that can be used as a planning tool and as a progress indicator. C) Create a PERT/CPM activity network for this project. D) Calculate the earliest and latest start times for each task along with slack times.

Activity	Est. time (hours)
1 Generate ideas	1.5
2 Build, test chassis, and report	4.5
3 Finalize parts list	1
4 Surf Web to find parts	3
5 Buy components	1.5
6 Study statement	2
7 Library research to find parts	2.5
8 Sketch ideas	3
9 Write progress report	2
10 Receive statement	0

2. What topics would you include in a well-rounded design proposal? What stylebook would you use?

3. Read the excerpted portions of a student design proposal on page 43. If you were responsible for accepting or rejecting this proposal what questions might you ask the designers? What additional information might you seek from them? How (if at all) might you suggest they enhance the organization of their design proposal?

4. Define consensus. How is a consensus-based decision different from a compromise?

5. How can you ensure open participation from your group? Avoid groupthink?

6. When (not if) conflict arises, how can you manage it most effectively? How do you characterize your own conflict management style? What if any implications do you think your style has had for your design team as a whole?

7. Although we describe decision making and collaboration as distinct skill sets, in terms of formulating design solutions how might they complement one another?

DORM ROOM DESIGN PROBLEMS FOR CHAPTER 5: FORMULATING SOLUTIONS (PROJECT AND PEOPLE SKILLS)

Assignment 10 (project management)

Having evaluated your ideas you should now be in a good position to plan out the remaining work you have to complete. Prepare a Gantt chart or Pert/CPM network documenting the work you have completed and the work that remains.

Assignment 11 (communication)

Organize a project proposal that explains your design recommendations and your work plan for completing it. Be sure to include the various elements described in Section 3.3.1.

Assignment 12 (collaboration)

Now is the time to take stock of your team's working relationship. Are you collaborating? Is everyone involved in the process? Are all ideas heard and seriously considered? Have you made sure your team members are not focused on a few ideas (groupthink) but have taken an open approach to the problem? Take a few moments and have people reflect on their individual feelings about the project and the team. As you look ahead to the next assignments, set up some roles for each of the team members. Each of you can take a turn at setting the meeting agenda. Each can take a turn taking minutes of what went on. Each can take a turn at managing the time in the meeting or work session to make sure the team stays on task. Write out a list of who will play which role in the next five meetings/work sessions of whatever type there will be. Answer the questions in the behavioral checklist on the following page and turn them in to your teacher.

BEHAVIORAL CHECKLIST FOR PHASE 2: FORMULATING THE SOLUTION

Instructions: List and then review each behavioral statement in the first column and reflect upon your experiences working with your current design team. Then use the scale below to rate your own effectiveness with regard to each behavior. Record your ratings in the second column. Use the third column to rate the effectiveness of your team. You may want to discuss your ratings with the rest of your team. In addition, you can make copies of this form and ask your team members to rate you as well.[1]

[1] Appendix A contains a development planning form you can use to establish improvement goals in relation to specific behaviors. Focus first on areas with the lowest ratings based on your self-assessment and/or any additional feedback you receive from team members or your instructor.

Rating Scale: 1 = Never 2 = Rarely 3 = Sometimes 4 = Frequently 5 = Always
N = Does Not Apply

Decision making	**You**	**Your team**
1. Identified several alternatives before selecting a solution		
2. Objectively evaluated alternatives in relation to criteria derived from functional requirements and constraints		
3. Avoided rushing to judgment on other's ideas		
4. Built on other's ideas and suggestions		
5. Encouraged unusual and creative ideas		
Project management	**You**	**Your team**
6. Thoroughly identified all tasks and activities that had to be completed		
7. Established realistic deadlines and time estimates		
8. Began documenting team progress and action steps		
9. Used a Gantt chart or other planning tool to track progress		
10. Modified tasks and timelines as new information was obtained		
Communication	**You**	**Your team**
11. Gathered information and research from multiple and relevant sources		
12. Clearly articulated how the proposed solution meets design requirements		
13. Appropriately documented references and resources used		
14. Used graphics and diagrams to illustrate points and ideas		
Collaboration	**You**	**Your team**
15. Involved others in discussions and decisions by soliciting their input		
16. Invited questions and comments from team members		
17. Encouraged contrary opinions		
18. Allowed time for debate and discussion		
19. Sought to ensure balanced participation among all team members		

BIBLIOGRAPHY

FENTIMAN, A. *Team Design Projects for Beginning Engineering Students*, Technical Report ETM-10-05-958. Gateway Engineering Education Coalition, Philadelphia, 1997.

JANIS, I. L. *Victims of Groupthink,* 2nd ed. Houghton-Mifflin, Boston, 1982.

MAIER, M. *Red Flags, Smart People, Flawed Decisions: Morton Thiokol and the NASA Space Shuttle Challenger Disaster*. SUNY-Binghamton, Binghamton, NY, 1992.

SCHWARTZ, R. *The Skilled Facilitator: Practical Wisdom for Developing Groups*. Josey-Bass, San Francisco, 1994.

THOMAS K. W., and KILMANN, R. H. *The Thomas-Kilmann Conflict Mode Instrument.* Xicom, Tuxedo, NY, 1974.

WEGGEL, R., ARMS, V., MAKUFKA, M., AND MITCHELL, J. *Engineering Design for Freshmen*, Technical Report ETM-1058. Gateway Engineering Education Coalition, Philadelphia, 1998.

CHAPTER 6

DEVELOPING MODELS AND PROTOTYPES (STEPS AND DECISION MAKING)

After a team has developed a concept for the design, members need to determine whether it will satisfy the design goals. They want to do this without manufacturing the final device, which might be expensive to produce. Instead, they might want to do a mock-up or simulation. This is the purpose of modeling, and it is one of the activities that differentiates engineers from people who rely entirely on experience or trial-and-error experimentation. Before describing the steps and skills associated with this phase of the design process, it helps to first define the term modeling.

Modeling is the process of representing your design concepts and abstract ideas without actually having to build the real thing. Models can come in a variety of forms. They can be sketches and drawings, physical representations, or computer images. They can be virtual replicas of what you expect your finished design to look like, or they can be abstract representations of physical properties associated with your design, such as load and force or horsepower and speed.

Regardless of the form and type of modeling used, to be useful all modeling must be done in an orderly and systematic manner so that tests can be performed, data can be collected, and your design's features can be enhanced.

There are four general reasons why engineers use modeling. The first is to develop a better understanding of the design problem and/or possible solutions. To some extent we already discussed this role for modeling (Section 2.1.3) when we described how engineering sketches could help you better understand design constraints and limitations. These sketches, or what are sometimes called *ideation drawings*, can be done on paper or on the computer. In any format they help engineers integrate information gathered about the design project with what they might have envisioned in their minds. By copying and refining ideation drawings, engineers can encourage new ideas and information.

The second reason engineers use modeling is to save time and money. Models enable you to gather information without incurring the costs (financial, time, and other resources) needed to make the final assembly. Note that here we are referring to two types of modeling. In the first type we determine the shape and size of the object (a descriptive model). In the second type we represent the object mathematically to determine the performance parameters for the design (a predictive model).

To gain information about the system that is being designed, the model should be constructed in such a way that the design variables can be changed to study their effects on all other variables. This brings us to the third reason why modeling is important. Specifically, it is through modeling that we are able to refine design ideas until we finally reach an optimal working solution. Modeling permits us to fix flaws continuously until we eventually reach a solution that meets design requirements and constraints.

A fourth reason for modeling is that it helps engineers convey ideas and information. By providing a visual representation of design concepts, models play a critical role in enabling engineers to communicate with one another and with clients, customers, and managers.

Before we proceed with a discussion of the steps in this phase of design you should also be familiar with the term *prototype*. This term closely relates to modeling but it is not exactly the same thing. A prototype is usually the first functional and full-scale representation of a design. It enables designers to demonstrate that their concept will work in the *actual* environment for which it was intended. The assistive feeding device and the autonomous robot projects discussed throughout this text are examples of prototypes. In fact, many student design projects result in the creation of prototypes because they are one-of-a kind representations. In the course of creating a prototype, engineers may use a variety of models and modeling techniques. On the other hand, prototypes are akin to modeling in that their purpose is to demonstrate feasibility and, in some cases (as with airplane prototypes), to continue testing a concept.

As indicated in Table 6.1, a large part of the modeling process is decision making and communications. It is important for individual teams and team members to be able to make decisions based on the results of the modeling process. It is also important for them to be able to communicate both the results of the modeling process and the decisions to their other team members, as well as to managers, clients, and other interested parties.

TABLE 6.1 Overview of Phase 3: Developing Models and Prototypes

Steps for developing models and prototypes	Skills and tools for formulating solutions			
	Decision making	Project management	Communication	Collaboration
1. Selecting a modeling process 2. Performing design analyses 3. Testing the overall design 4. Revising refining, and critiquing the design	• Quantitative and qualitative decisions • Conducting design and critical reviews	• Clarifying roles and responsibilities • Obtaining resources	• Writing progress reports • Providing feedback • Seeking input and feedback	• Managing role conflict and role ambiguity • Recognizing style differences • Eliminating social loafing

The steps in the modeling process are as follows:

1. Selecting a modeling process
2. Performing a design analysis
3. Revising, refining, and critiquing the design
4. Making decisions based on the modeling and analysis.

In this chapter, we focus on the steps and the decision-making skills that go along with executing them. We explore each of these steps separately in the following section. As you read this chapter, keep in mind that *selecting* a modeling process and performing a design analysis are really two sides of the same coin. You have to know what kinds of analyses you want to perform before you can determine what modeling process to select. By the same token, you have to understand the strengths and limitations of various modeling processes in order to know if they will meet your needs.

In addition, keep in mind that decision-making during this phase can best be characterized as critical evaluations of procedures and results. When making these decisions it is important to remember the specific criteria and functional requirements of your design. It is against these standards that you will want to interpret test results and critique the quality of your design.

6.1 SELECTING A MODELING PROCESS

To select an appropriate modeling process, you have to be familiar with a whole array of processes. Therefore, in this section we describe various modeling processes and provide some examples of their applications. In the broadest sense, all modeling processes can be grouped into two categories. *Descriptive models* are those that depict ideas, products, and processes in a way that is recognizable. As the name implies, the goal of these models is to show what a design would look like if it were created. Examples of descriptive models are engineering drawings or three-dimensional computer models. They could be depictions of an entire project, such as an architect's model of a building, or representations of some smaller part within a larger design, like a diagram of the CD-ROM drive in a computer. In either case, descriptive models are an important ways to communicate your design ideas, but they cannot be used to predict the performance properties of your design unless you are trying only to assess aesthetic characteristics.

Predictive models, on the other hand, are those that can be used to test and understand how design ideas, products, and processes will perform. In some cases, predictive models bear little or no physical resemblance to the overall design. For instance, some predictive models are nothing more than mathematical equations or graphs showing relationships between design requirements (e.g., lift-and-drag properties of an airplane). In other instances, predictive models also have descriptive qualities.

Scale models are one of the most basic and generally useful types of models. They are essentially physical representations of a design. Depending on what you are designing, scale models can be full size or replicas made to scale. There are numerous types of scale models: some are just descriptive applications and others are used for predictive purposes as well.

You probably have seen examples of scale models. For instance, architects build a scale model of the building they are designing. In the chemical processing industry, designers build scale models of an entire plant. In this case they might use scale-model plastic pipes. In the latter case, some models are equipped with miniature vehicles that are "driven" through the model to check for clearances (a predictive application of scale model). Civil engineers, landscape architects, and city and regional planners often build models of the terrain they are modifying. Electrical engineers assemble electrical and electronic components on a circuit board called a breadboard.

Students have a surprisingly broad range of scale modeling processes at their disposal. Obviously they can use sketches to model their ideas. Students also can build mock-ups with poster board, chipboard, or cardboard that takes them from the sketch to a *physical model* they can examine, touch, and explore. Students can also use plain paper, tongue depressors, swizzle sticks, and plastic utensils.

Using Erector sets and Lego sets are additional ways to make scale models. Today the Erector sets and Lego DACTA models include motors, gears, pulleys, chains, and sprockets. The new Lego kits provide a controller and sensors in addition to the drive train components, which allow the models to be animated. A very simple Lego model of a linkage is shown in Figure 6.1. With this model, it is possible to determine the motion characteristics of the linkage and even to gain some appreciation for force transmission.

All these methods are very much hands-on and can help give students a good feel for their design and its features. However, because they are hands-on, these methods also make it easy to approach problems through unstructured trial and error. For instance, when some student teams first get their hands on an Erector set or a Lego set, they try to build the first idea that comes to mind. As a result, they frequently pursue a fruitless path and are reluctant to make changes until they eventually run into a roadblock and have to start all over, this time with a plan.

Maintaining a more systematic, engineering approach to design modeling enables you to take full advantage of the time saving and learning that modeling can provide. The engineering approach calls for a team to explore possibilities, compare the possibilities, make appropriate analyses, and then choose the one to pursue.

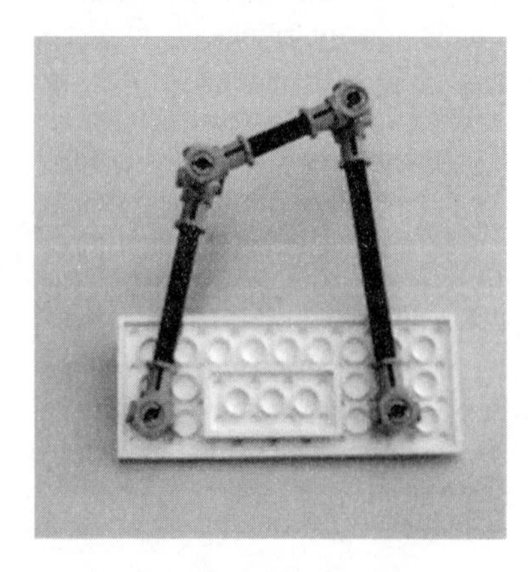

FIGURE 6.1 Simple Lego model of four-bar linkage. (Courtesy Gray Kinzel.)

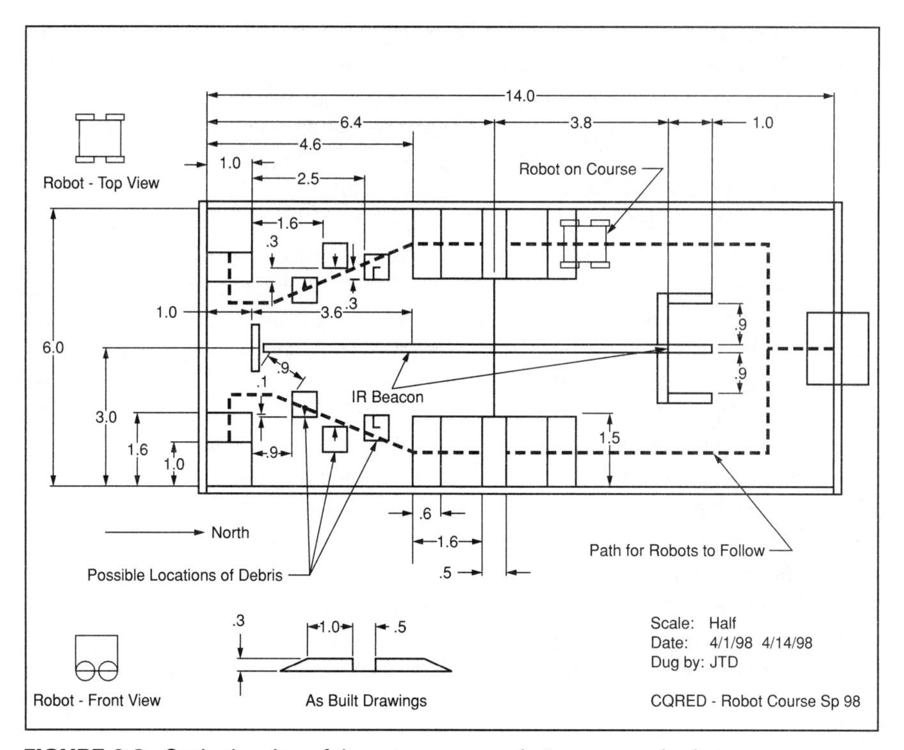

FIGURE 6.2 Scale drawing of the autonomous robot course and robot.

The level of sophistication of the systematic approach depends on the level of the project. At the freshman level, the analyses may rely primarily on geometry, elementary mathematics, and/or scale models. At the senior level, students should be able to use complex analysis techniques and software at a performance level near that expected of a graduate engineer.

A brief case description illustrates what we mean by taking a systematic engineering approach to using a scale model. When student design teams are working on the autonomous robot problems, they are urged to make a scale drawing of the course plan view and a scale drawing of the plan view of their robot (see Figure 6.2). This can include making a drawing of the ramps on the course in both plan and elevation (side) views. They can then drive the robot around the course to see where it may have problems turning, climbing, descending, picking things up, and so forth. They can test whether the front or back (or sides) of the robot will drag when it attempts to go up or down a ramp.

The robots will have a time limit in which to complete the course. The teams measure the path on the course to determine the number of feet or meters that must be covered in the time limit (typically 2 minutes). They can then develop the minimum speed at which the robot must travel. When they allow time for turning, getting lost, and picking up things, they then have to change their estimates of minimum speed. The paper model in this case provides considerable information about shape and size as well as potential problems without much expense and without taking a lot of time.

The ever-increasing sophistication of computer technology has made *geometric modeling* a common tool as well. Today most computer-aided design and drafting (CADD) systems provide three-dimensional drawing capabilities. Many of these programs also now include "walk-through" capability so the viewer can follow a path through a house, across terrain, or through a system. Some popular CADD programs are CADKEY, AutoCAD, ProEngineer, Solidworks, SolidEdge, IronCAD, and Silverscreen. By using them students can quickly modify various parameters and features to systematically assess the pros and cons of various design features. Three-dimensional geometric models developed on a CADD system are either wireframe, surface, or solid models. *Wireframe models* can be used as input geometry for simple analysis work such as kinematic analysis and finite element analysis (see Section 6.2). *Surface models* can help you visualize your design by allowing you to add things like color and shading. *Solid models* are mathematically accurate descriptions that can be assigned different proportions and properties (e.g., material and texture). As a result solid models can be used for both describing and predicting. Because solid models are extremely realistic, they are frequently used for *rapid prototyping*. Rapid prototyping is the generic term for several related processes that create one-of-a-kind physical models directly from the parameters specified in a three-dimensional CADD program. It is frequently used to create physical models of industrial and consumer products.

Although few student design projects involve rapid prototyping, they commonly use geometric models, especially solid models. As an example, consider the assistive feeding device project. In this case solid models were crucial as both descriptive and predictive tools. To develop the solid models, students at the various universities involved in the project primarily used SolidWorks and ProEngineer. Both programs are *parametric*, which means that the information for the model is stored in equation and constraint format instead of in a simple numerical format. This allows the user to change individual features in a part of the assembly, and the programs then update all other features that depend on the ones that are changed. This capability is essential in any solid modeling system considered for student projects. Fortunately, most of the current versions of the popular solid modeling packages are based on parametrics. With solid models, the students detected a number of problem points in the robot arm. Places were discovered where it would be difficult to insert bolts, and, in several places, interferences were identified. A solid model of one version of the wheelchair and arm assembly is shown in Figure 6.3.

With solid models, the students were able to model the components of the robot arm and to generate an accurate geometric representation. The solid models were the basis for most of the future analyses and data sharing. The students at different universities were also able to share information with each other by sending attached e-mail files or by posting the files on Web sites from which they could be retrieved with a file transfer protocol (FTP). FTP worked best when the students used the same programs to read and write the files; however, to a limited extent students were able to translate files from one program to another. For example, it was possible to use SolidWorks to display a ProEngineer file, but it was not possible to edit the file in SolidWorks.

Students working on the robot arm also used *mathematical models* extensively. Although they are abstractions, mathematical models are very useful for understanding and predicting performance, especially that of large, complex systems that often

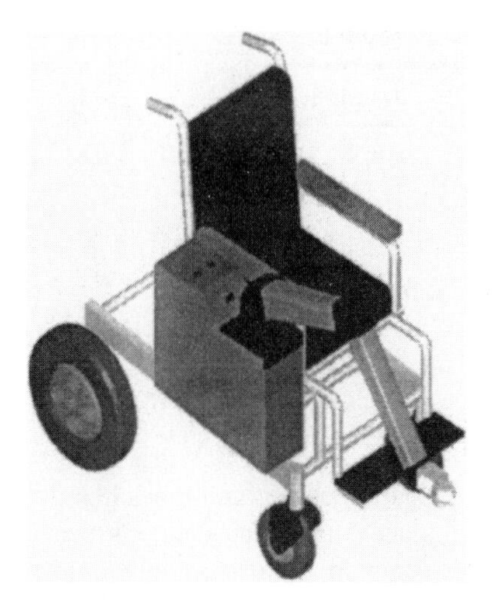

FIGURE 6.3 Solid model of design concept for robot arm on wheelchair.

are impractical to test in physical form. It is far less expensive to change numbers in a computerized mathematical model than it is to make physical changes to a mechanical prototype.

For instance, students working on the robot arm needed to determine whether the arm was strong enough to support the specified loads and to withstand unexpected loads. The students were required to compute the stresses in the major components. Long, straight components could be modeled as beams that would be analyzed with the equations learned in a strength of materials course. These equations were programmed in MATLAB and were used to optimize the linkage arms for weight and to determine the suitability of using standard structural shapes. When it is possible to make such calculations, they can normally be made in a matter of minutes, and the equations employed can be used as design equations. This means that physical parameters such as a wall thickness or beam height can be easily changed to determine the effect on the final stress values. Structural shapes that were not simple beams or shafts were analyzed by the finite element method. These were members that had complex cross sections, cross sections that changed abruptly with length, or a centerline that was not approximately straight. The programs used for the analysis were COSMOSWorks and ANSYS. Both programs accept a solid model and automatically generate the finite element mesh. Therefore, the modeling procedure is relatively simple: generate a solid model, save it in the appropriate file format, and read the file into the finite element program. Considerable engineering judgment is often required, however, to interpret the results and to ensure that the results are reasonable. To determine whether the results are reasonable, it is desirable to make "ballpark" calculations with simple strength-of-material equations.

The last two types of modeling you should be familiar with are *computer simulation* and *computer animation*. Computer simulation involves using precise three-dimensional computer models to test a design. For instance, if we want to test how well certain materials will work in a design (for property analysis, see the following section) instead of subjecting a physical model to external loads, we can

TABLE 6.2 Categorization of Engineering Design Models

Design models	Predictive	Descriptive
Scale	X	X
Geometry	X	X
a) Wireframe	X	
b) Surface		X
c) Solid	X	X
Mathematical	X	
Computer simulation	X	X
Computer animation		X

assign material properties to a computer model so that it behaves and looks like the real thing. In other words, the computer model simulates the real product or system. The finite element models discussed before are an example of this. As another more specific example, instead of placing a scale model of an aircraft in an actual wind tunnel, a computer model could be used to simulate what would occur during the test. This includes precisely representing the properties of the aircraft and the fluid properties of the air. A sophisticated simulation program is required but, fortunately, such programs are available to students for most of the types of computer simulations that are of interest on typical design projects.

Computer animation, on the other hand, is an imprecise modeling of a design, situation, or characteristic, although in some situations the animation can be somewhat precise. When the animation is fully precise and accurately represents the real situation, then it has become a form of what we have already described as computer simulation. Thus, an animation is realistic, but otherwise imprecise, compared with a simulation, which is both realistic and precise in its replication of a situation. As an example of the distinction between these two, a computer animation could be used to present a simple *visual* representation of an aircraft in flight, but a computer simulation would be required to properly determine the aerodynamic *performance* characteristics of an airplane in flight. Computer animation is most valuable when the situation to be analyzed involves time or the passage of time as an important element.

Now that we have reviewed the various types of design models that engineers use, the next topic to discuss is the kind of analyses engineers perform. Before we do that, however, take a look at Table 6.2 for a quick review of dsign models and an indication of whether they can be used for descriptive purposes, prescriptive purposes, or both.

6.2 PERFORMING DESIGN ANALYSES

As we mentioned at the beginning of this chapter, modeling is the process of representation without actually building a final version of a design. We create models so that we can perform design analyses. Therefore, the main question to ask in order to select a modeling processes is "What kinds of analyses do we need to perform?" To perform a design analysis means evaluating a design to see if it meets the criteria articulated in the defining the problem and formulating solutions phases.

Table 6.3 provides a brief description of some of the most common types of analyses engineers perform. It also lists typical software tools used to perform these analyses. As your engineering education progresses you will learn more about how to perform these analyses and interpret results from them. For now our goal is to make you aware of what they are and how they are used.

We use examples from the autonomous robot project and the assistive feeding device project to introduce you to various types of analyses. In each project, the goals for the analyses were to model the system to determine how large to make components to avoid failure and to determine the appropriate geometry that would allow the devices to satisfy functional requirements. Another more subtle reason for the models was to gain a general insight into the performance of the systems. This helped the students to predict what things might go wrong and to model the effect of the changes required to fix designs when unforeseen problems occur.

Autonomous Robot Example In the robot design and construction, there were a number of components to be analyzed. The chassis must be rigid and the drive train must provide adequate power for each phase of the course limits. Functions such as gap crossing, object collection, and lifting must be analyzed to be sure that sufficient motion and power are available. Thinking about analysis can cause maneuvering to be overlooked. Teams sometimes forget to make sketches of the robot in all parts of the course to see whether the sensors are located properly. First- and second-year students normally have studied physics of mechanics or have studied it in high school. A number of different analyses that need to be made in many problems can be done by using the physics of mechanics or physics of circuits. If an object with simple supports is to carry a load, the supports for the object can be determined by using force balances (dynamic analysis).

If an object must move up a slope as the robots are required to do, then a simple static force balance can be performed. A dynamic force balance is no more difficult as long as acceleration is not involved. Determining the required power follows the force balance. Physics books and engineering mechanics of statics and dynamics can provide information and examples about how to perform these very basic types of *dynamic analyses*. Modeling can be done with a software program such as interactie physics or working model once the force balance is done.

A portion of the analysis can also include making a free body diagram and then doing a graphic vector solution. In such cases, the vectors can be drawn to scale and the resultant vector can be measured with an appropriate scale.

Assistive Feeding Device Example In the case of the feeding device, it was necessary to conduct a *dynamic force analysis*, a *kinematic analysis*, and a *property analysis* to evaluate stress. Mathematical models had to be developed to perform these analyses. Mathematical modeling was particularly important because the project's scope and complexity made it costly and impractical to make repeated physical changes to the prototype being developed.

At the beginning of the project, the sizes of the motors to actuate the joints of the robot were unknown. Determining the motor torques required developing free-body diagrams of every component in the system and then writing and solving equations of

TABLE 6.3 Descriptions of Engineering Design Analyses

Type of analysis	Description	Examples of engineering software tools that facilitate analyses*
Property analysis	← Evaluates physical properties such as strength; size; volume; center of gravity; weight; center of rotation; thermal, fluid or mechanical properties.	← *ANSYS, Algor* (finite element programs) ← *Fluent, FIDAP* (fluid analysis programs) ← *Interactive thermodynamics* (thermodynamics program) ← *HYSYS* (process simulation program) ← *HTFS* (heat transfer simulation program) ← *Flarenet* (flare system simulation programs) ← *Flarenet* (flare system simulation programs) ← *BioPro Designer* (biochemical simulation programs)
Finite element modeling	← Used to determine the static and dynamic responses of discrete design components under various conditions like different temperature, interaction with fluid flow (e.g., air, water)	← *ANSYS, ALGOR, COSMOS, NASTRAN, MARC* (general finite element programs)
Mechanism analysis	← Specifies motions and loads within mechanical systems made of rigid bodies connected by joints (e.g., a clamping device)	← *Working Model, Analytics, ADAMS* (kinematic and dynamic analysis of mechanism)
Assembly analysis	← Defines individual rigid bodies within a mechanism by considering both geometry and velocity (e.g., proper clearance required for adjoining parts in an engine)	← *ProEngineer, Unigraphics* (solid modeling programs with interference checking)
Kinematic analysis	← Determines range of motion of assembled parts without regard to load (e.g., a door hinge)	← *Working Model, Analytics, ADAMS, ProEngineer, Unigraphics* (for motion analysis of mechanisms)
Dynamic analysis	← Determines how loads drive or create the motion of a mechanism (e.g., how a hinge operates when attached to a door)	← *Working Model, Analytics, ADAMS* (kinematic and dynamic analysis of mechanism)

Functional analysis	← Evaluates the worth of a design in terms of issues like cost, appearance, usability, profitability, safety, marketability	← Excel (spreadsheets to compare the weighted values of different options)
Human factors analysis	← Identifies how a design interacts with its users in terms of physical and psychological characteristics like mental abilities, sensory perception, height, range of motion (e.g., determining what color lights work best on a flight instrument panel in a cockpit)	← *ADAMS* (kinematics program to simulate body motion)
Aesthetic analysis	← Evaluates the design based on its look and feel (e.g., making sure an automobile has a look that appeals to people in its target market)	← *SolidEdge, SolidWorks, ProEngineer, Ideas,* Unigraphics (solid modeling programs with realistic rendering capabilities)
Market analysis	← Analyses typically done even before a design is sold or produced to determine the needs and wants of customers	← *FileMaker Pro, MS Access* (database programs)
Financial analyses	← Analyses typically done even before a design is sold or produced to determine if there is sufficient capital to cover development expenses and whether the designing organization(s) can expect a return on the investment	← Excel (spreadsheets to compare the weighted values of different options)

*Only the program names are given here. A large amount of information on these program and others is available on the World Wide Web.

statics. Because of the serial nature of the device, it was relatively straightforward to analyze each link, starting with the hand and working toward the shoulder. With this approach, it was possible to predict accurately the forces at each of the bearings and to compute the torques the motors must supply. The motors could then be selected and the analysis could be completed with the weights of the motors included. To predict the effect of changes in the components, the force analysis equations were programmed with MATLAB.

Although determining the power and torque required was relatively straightforward, finding the motors was a difficult task. The motors not only had to deliver the torque required at a given speed, they also had to be small enough to fit into tubes used for the arm. The steps taken to determine the appropriate fit are an example of an *assembly analysis*. Executing this analysis required creating a specific type of mathematical model called a parametric model. An accurate parametric model of the tubes was necessary to ensure that the motors available would fit. As a result of this exercise, the design team determined that they needed to use a different tube and different material to accommodate the motors that were available. This is an example of design tradeoff, which is common in design. A tradeoff analysis is greatly enhanced by accurate mathematical models.

Kinematic analyses were used to determine how the arm would move. Specifically these analyses helped determine the reach of the arm and the general workspace for the robot hand. The position of the components is also necessary for a complete force analysis.

The positions of the components of the system initially were analyzed by two approaches. In one approach, the team developed the equations defining how the links would move given specific values for the rotation angles at the joints. For this, they used the minimum amount of information necessary to characterize accurately the motion of the arm links. This reduces the arm to the stick figure in Figure 6.4. The stick figure represents the rigid body in the structure and permits all the equations for the essential kinematic information to be computed. These equations were tedious to develop but were based mainly on trigonometry and geometry. The students programmed these equations with MATLAB and then animated the linkage and displayed graphs of various parameters on the computer screen (Figure 6.4). Both positions and velocities were conveniently modeled. This approach to kinematic modeling gave the students considerable flexibility to study the effects of different parameter changes. The kinematic analysis allowed them to calculate how different parts of the design move relative to each other and to animate the results. This was very educational for the students because they had to understand the problem thoroughly before they could program it.

A second approach to modeling the arm was to use a popular kinematic and dynamic software program, WorkingModel. This program requires very little user understanding of kinematics, and the user interface makes it very easy to model fairly complex kinematic structures. It also reads the geometry files directly from solid modeling programs such as SolidWorks and SolidEdge. Therefore, it is possible to generate very realistic animations of the arm with this program. Dynamic characteristics could also be modeled when the joint inputs were specified.

These analyses illustrate the modeling philosophy of considering only the detail necessary to determine the quantities desired. Based on the kinematic analysis and the

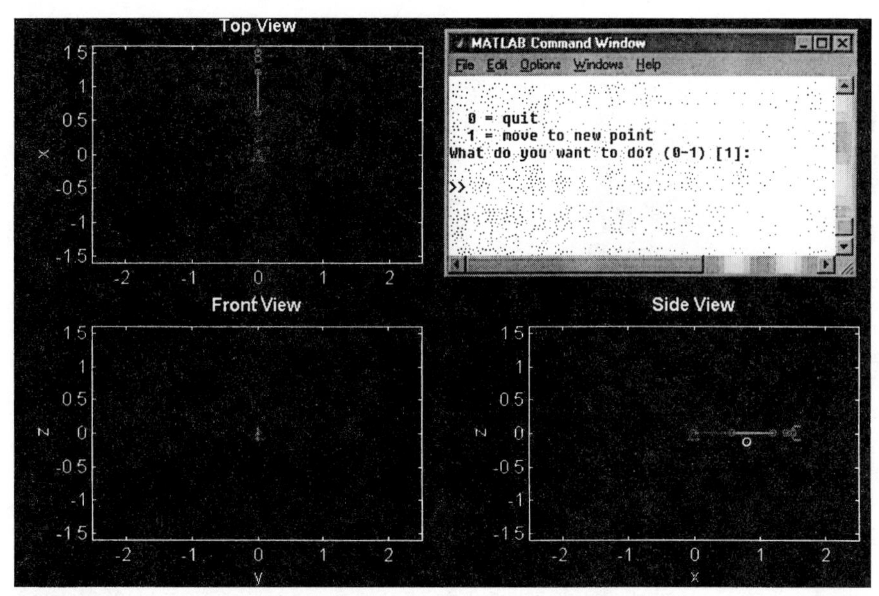

FIGURE 6.4 Initial position of linkage in Matlab system showing the use of stick figures to model.

force analysis, all the major mechanical components can be selected. These include the structural components and drive components such as bearings and gears. As individual components are selected, the components can be added to the model and the analysis refined. Students program the equations as they are developed because refinements can then be made fairly easily.

Property analyses were used to help design the brackets. It was not initially known how thick to make them or how much material could be removed from the cross sections to reduce weight and maintain adequate strength and stiffness. Furthermore, the shape of the brackets was (sufficiently complex that the stresses could not be computed accurately with simple strength-of-material equations. The brackets were refined iteratively by finite element modeling using COSMOSWorks. The thickness was selected by considering only standard sheet thicknesses. Next, the cross section was refined by removing material and reanalyzing it to determine the maximum stress levels for the worst-case loading conditions. The final acceptable geometry was established when the maximum expected stresses were essentially equal to the design stresses. The design stresses were determined by using the yield stress for the material and dividing it by the safety factor. For this design, the students used a safety factor of 2. The progression of geometries considered for the elbow bracket is shown in Figure 6.5.

The final part of the project requiring modeling is the control system. It is extremely important for the control system to be modeled accurately before the hardware is designed because of the expense associated with the components as well as for safety considerations. The control system is the heart of the robot, and it will not function properly if the control system is faulty. One of the features of the project

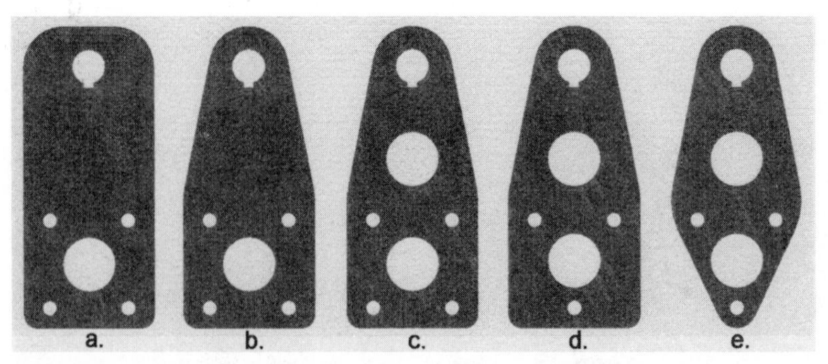

FIGURE 6.5 Elbow bracket design evolution for robotic arm.

is a game input device to orient the hand of the robot without having to actuate the joints of the arm individually. The computer model required coupling the control software with the kinematic analysis equations. The students did this with a combination of custom programming in C and the MATLAB control system simulation toolbox SIMULINK.

On a complex project, it is necessary to model every major component. It is not normally possible simply to draw, manufacture, and test a system successfully without major problems unless the designers have extensive experience. In the assistive feeding device project, a major problem resulted when the students did a quick mental evaluation for the size of one of the elbow motors. The resulting motor was twice as large as needed, and it reduced the range of applicability of the design. Kinematic and dynamic analyses could have identified this problem. Specifically, a careful engineering force analysis using simple statics would have revealed the problem before the motor was ordered and the arm was manufactured and tested.

Most of the analyses we have discussed so far have focused on determining whether components move correctly, are large or small enough, and are strong or stiff enough. All these are determined by quantitative or predictive analyses. There is also *qualitative or aesthetic analysis*. How do things look? Are they pleasing? Do they present the right image? Do they feel right?

Again, in *All Corvettes Are Red*, the company worried about whether the new design would carry the evolving lines of the previous Corvettes. They wanted the new machine to be instantly recognizable as a Corvette but also as the latest Corvette. When the qualitative analysis was complete and the body was approved, the engineers started to work on the engine, chassis, controls, seating, heating and air conditioning, and so forth. Because the body had been done first, the engineers had to try to make things fit in the spaces provided and they had to go back to the body design group for changes and discussions about how to solve space problems. Quantitative design and analysis came after the qualitative analysis.

At one of the leading car manufacturers, an engineering summer intern was assigned to determine whether a car roof felt strong enough. To do this he assembled a group of people within the company from different backgrounds and had them feel the roofs of a variety of the company's cars and of competitors' cars. The people had to report which roofs felt strong or weak, and the summer intern summarized the findings and made a report. The engineering team working on the new car design then studied the acceptable roofs, made appropriate measurements, and designed and analyzed a roof that provided the proper feel.

In general it helps to start with simpler analyses and then proceed to more complex analyses. For instance, in the above example, the design team first obtained a qualitative understanding of what strong meant before they tried to design and test a roof.

6.3 TESTING THE OVERALL DESIGN

Analyzing various subcomponents of an overall design helps to ensure that we can understand unique characteristics of the various parts and subsystems within our design. This way we obtain a complete appreciation for what and how things work. In complex projects, it also enables us to prioritize work and to delegate different tasks to various team members. However, understanding how things work independently is only part of the story. We also must understand how design components work in concert with one another and eventually systematically reach a point where we are able to test the overall design as a complete unit.

In the robotic arm project, the students developed a complete solid model from the parts supplied at each of the participating universities. This provided a general check on whether all parts would work together. In addition to this, physical models were constructed of some individual components. For example, one of the original gripper designs was first created in plastic to determine the motion and also to gain a physical feel for the overall size and ease of operation. Physical models are extremely useful for investigating how easily different components will move relative to one another. Often, a solid model indicates that various parts work well together, but a physical model reveals that the parts bind because of friction.

6.4 REVISING, REFINING, AND CRITIQUING THE DESIGN

As the design progresses, it is necessary to evaluate the different aspects of the design to determine whether there are any factors that might prevent the project from being a success. This means critiquing the design to identifying weaknesses and then revising the design to eliminate problem areas. The initial objective is for the design to satisfy the performance objective and to make any changes necessary to ensure that the objectives are met. Then, if there is time and money left in the project, the design can be refined to make it conform to criteria that either the designer or the customer specifies.

Autonomous Robot Example For the students working on the autonomous robots there was a lot of revising and refining. The critiquing part was stated generally: "Will it travel through the course and do what it is supposed to do?"

The revising and refining for the student teams incorporated everything from adjusting things that were designed properly to rebuilding the robot from scratch.

Some examples of refinement dealt with things such as adjusting the belt tension in both the tracks so that the robot would go forward in a straight line. This normally had to be combined with adjusting the power setting for the motor ports so that the drive motors were rotating at the same speed. The teams found that the same power setting for two matched motors did not produce the same speed in each motor.

Refinement also dealt with motor power settings and motor "on" times so that the robot would turn the corner at a proper angle. As the students worked with their robots, they found that the "freshness" of the battery charge could be critical to things such as turning and speed to complete the course.

One team had pursued a particular design for their robot through the eighth week when the preliminary competition was held. The robot had performed adequately in the weekly performance reviews. However, their robot ran so poorly in competition that they built a new chassis and drive train and reprogrammed the controller for the system. In the ninth week, at the final head-to-head competition, they finished somewhere in the middle to upper end of the teams. Obviously, in the process of building the first design, they had learned a lot about what worked and what didn't.

Cost is also a factor for the teams. As they continue to refine and perhaps add components, they must keep track of the total cost. Teams can earn extra points for a low-cost design. If their effort comes in under budget, they can add points to their score. If they come in over budget, points are taken away from the final score.

Assistive Feeding Device Example As the robot arm project evolved, compromises in the design were necessary. To meet safety and weight requirements, it was necessary to place the elbow motor inside the tube for the lower arm link. However, torque requirements dictated that the diameter of the motor be approximately 2 inches. Originally, the tubing was to be made from fiber-reinforced plastic, but the students were unable to find tubing large enough in cross section. Therefore, they changed the material for the arms from plastic to aluminum.

In another area, the thrust bearing selected for the base was too small. It was selected for only a normal load, but the primary load on the bearing is from the movement produced by the arm and payload. This was not recognized until the arm was tested in the final assembly. This is somewhat typical of the problems associated with prototype development. It is extremely difficult to account for all the problems that might be indicated until a physical prototype is developed. Regardless of the care used in the modeling phase, it is important to be prepared to make changes and refinements after the device is first assembled. It is a rule of thumb in some development organizations that half the project funds should be available for refinements after the prototype is first assembled and turned on.

Conducting design/critical reviews are one way to minimize the amount and extent of revisions that must occur on a completed or close to completed design. Design reviews are structured meetings during which engineers give or receive feedback about an engineering design or concept. They are common and critical phases of virtually all engineering design projects. The goal of design reviews is to provide a forum for critiquing designs in a manner that helps engineers identify ways to improve their design solution before it is completed.

Your instructor may have built periodic reviews into your design process. If so, this section will help you to be better prepared for those reviews and obtain the most out of them. Whether reviews are a formal requirement for you, it is a good idea to conduct them with your team members anyway. Try to conduct at least two design review sessions with your team members in addition to any reviews your instructor may require. A good time to hold the first review is after you have formulated your initial design solution. A good time to hold the second review is when your team is close to completing your model or prototype. Try to get outside people to participate in the review. These may be classmates or perhaps outside experts from industry. A good review board has three to six reviewers. Keep the design review focused. Identify specific parts of your design about which you want critiques and feedback You should make a checklist of concepts and features that the review will focus on. To create this list refer back to your design's functional requirements.

Avoid lapsing into a problem-solving/design mode. This is a time for reviewing and critiquing and not necessarily for developing new ideas in detail. Keep notes of the issues raised and plan on summarizing key points with participants before the meeting ends. In addition, you will want apply the principles of giving and receiving constructive feedback that are discussed in Chapter 7.

Keep in mind that it is hard for most of us to separate our sense of ourselves from our work. For this reason, design reviews can be a blow to our egos and have the potential to leave us feeling pretty beat up. Try not let yourself fall into this trap. Focus instead on the value and wealth of information a design review can provide. Still, feeling intimidated by the design review process, especially when it is new to you, is natural. One very successful engineer who later went on to be a senior officer of a Fortune 500 organization recounted the sense of deflation he felt when he left his very first design review. "I went into that meeting feeling proud of my efforts and my results, but by the time it ended I wasn't sure I could pass a high school algebra test. Imagine my surprise," he continued, "when after the meeting my boss met me in the hallway and grinning profusely, congratulated me on how well the meeting had gone!" The point is design reviews, by their nature, are meant to be critical, but the information they provide is invaluable to developing successful models and prototypes.

Using Design Reviews in Student Projects

Autonomous Robot Example

When the student teams are working on their robots, they have a weekly presentation to the faculty. In each of these reviews, the students get back previously written materials with grades and notes, make an informal oral presentation of work accomplished in the past week, and demonstrate that their design can meet the weekly performance review requirements. The team can raise any issues or concerns with the faculty. When the competition is complete, they have to prepare an oral presentation and give the presentation to their instructor and their classmates. They also have to provide a written report and a project notebook.

The teams are required to do the final report outline by midterm, do drafts of the first half of the report by 2 to 3 weeks before the end of the term, and do a draft of the oral presentation outline a week before the end of the term. In each case, the faculty reviews the outline or draft, makes corrections, return the materials, and goes over the corrections and suggestions in the weekly reviews.

Assistive Feeding Device Example

The assistive feeding device project is structured so that a formal review with the faculty advisors is scheduled about two-thirds of the way through the project. It is typically a 7-month project, and the critical design review is scheduled for the fifth month. The students are required to present their parts of the project and to have a rough physical model or prototype for the components. This allows all the participants to view the project as a whole and to determine whether there are any obvious problems with individual components. Minor problems can be resolved at the critical design review meeting, and major problems are resolved between the time of the review meeting and the final meeting.

Formal reviews are critical to the success of a project such as this one. Deadlines are significant motivators for students, and even though the students from the individual schools are working on the same team, there is always a sense of wanting to represent each university well. Therefore, the students make an effort to achieve the goals set for each deadline. We have found that, when we do not set the individual deadlines, the students and faculty tend to put things off so that there is far too much to do at the end to produce a quality product.

CHAPTER REVIEW

Testing and modeling are very much at the core of what engineers do. During your education and engineering careers you will learn more about the specific types of modeling methods and analyses that are most relevant to your engineering specialization. Our goal in this chapter has been to provide you with an overview of what modeling is and how it helps engineers. Engineers rely on models to obtain information, provide information to others, save time and money, and generally develop a more detailed understanding of their ideas. You should recognize that there are two broad categories of modeling techniques: descriptive and predictive. You should also appreciate the fact that engineers conduct a wide variety of analyses. These range from relatively subjective investigations of qualitative and aesthetic design characteristics to detailed investigations based on abstractions and data that in and of themselves bear little physical resemblance to the eventual design.

REVIEW QUESTIONS

1. What are some of the main reasons why engineers use models?

2. How is a model different from a prototype? How is it similar?

3. Explain what is meant by the terms predictive modeling and descriptive modeling. Give examples of each.

4. Wireframe, surface, and solid are all examples of what kind of modeling? What are the differences between them?

5. When might I want to use computer animation as a modeling technique? When might I want to use computer simulation?

6. List some of the benefits and limitations of doing mathematical modeling. Provide two examples of types of analyses that usually require mathematical models.

7. What type of anlaysis would you perform if you wanted to determine how two subcomponents of a design fit together?

8. What are some reasons for testing subcomponents of a design separately from one another? Why do we still need to test the overall design as a unit?

9. If you were to conduct a design review meeting what are some specific things you might do to ensure that the meeting was useful?

DORM ROOM DESIGN PROBLEMS FOR CHAPTER 6: DEVELOPING MODELS AND PROTOTYPES (STEPS AND DECISION MAKING)

Assignment 14 (decision making)

Draw the plan view of the room and an interior elevation of each of the walls. Use grid paper that allows you to show the appropriate detail for the dorm room. If available, pick a size such as 11 × 14, 11 × 17, or 17 × 24 feet. Depending on your design, you may also want to do a view of the ceiling as well. Now draw both the plan and profile views of the furniture and other objects that must be stored in the room. You can then move the drawings of the furniture around on the views of the room to determine spaces left and size. Is it possible to keep your bike in your room?

Assignment 15 (decision making)

Carefully study each of the elements that will have to carry loads of any type as part of your designs. Where are loads applied? Make a list of these and determine what type of loads (point or distributed) are applied. Think about what a student might put on each of these "hangers." Would someone grab a hook attached to a wall or door and hang on it? Would they hang a book bag or a bike on it? If the student invites several friends in, is it likely that three or more might sit on the bed? If so is your bed strong enough to hold three or four students? Remember that some loads may be dynamic. For example, students could hold their book bags above a hook and then let go. The students could also use a bed for a trampoline! Once you have made the list and described the loads, determine which problems you are ready to solve and which ones will require more instruction from your professor or more research and study in the physics or engineering mechanics books. It certainly is possible to make estimates and then double your estimate to provide safety. Whereas some tools, like a computer, provide a concentrated load or point load, other objects like a mattress provide a distributed load. On a rod in the closet, the clothes provide a series of point loads that may be estimated as a distributed load. You need to be concerned about the bar and hooks being strong enough to hold the load that may be attached.

Assignment 16 (decision making)

It is now timed to create a scale model of the room you have designed. The furniture you have chosen for your room can be made with a combination of poster board, toothpicks or small wooden dowels, and fine wire. Your team may find that modeling clay works well for making small objects. The wooden dowels, poster board, and wire can be used to make a bike if you plan to store a bicycle in your room. Markers can be used quickly to provide color for the various components. Move the furniture around to see if you have created the optimum design. You should have been carrying along some alternatives to what your team believes is the best design. Make models of these components and compare your alternatives. Consider how your model can be used to test for compliance with important local, state and federal requirements. For instance, how are you going to test for wheel chair access, something that is required under the Americans with Disabilities Act.

Assignment 17 (decision making)

From the testing that was done as part of Assignment 16, you probably found some things that did not work. Before you make the changes, gather a group of your classmates and have them review your designs. When you have their suggestions and your own ideas about getting the best solution, make the changes to your design components. Don't forget to document what you learned from the design reviews that your team and your classmates provided. Whether you are looking at a chosen design or at some options, it is time to consider the cost of your dorm room components. If you are doing a hundred rooms, a small increase or decrease in cost can multiply the total cost rather quickly.

BIBLIOGRAPHY

BERTOLINE, G., WIEBE, E. N., MILLER, C. J., and NASMAN, L. *Technical Graphics Communication.* Irwin, Chicago, 1995.

SCHEFTER, J. *All Corvettes Are Red.* Pocket Books, New York, 1996.

CHAPTER 7

DEVELOPING MODELS AND PROTOTYPES: PROJECT AND PEOPLE SKILLS

You probably realize from the previous chapter that much of what you will do during this phase is based on analysis and can be highly technical in scope. However, this phase of the design process also involves a great deal of team interaction, coordination, and communication. Therefore, this chapter focuses on the tools and techniques for managing the project and people issues associated with developing models and prototypes (Table 7.1). In terms of project management, emphasis is placed on clarifying the specific work and roles associated with development and testing of your design. Oral and written communication remain important during this phase as well. Specifically, you should understand how to document your progress and test results. In addition, because critical evaluation is so important during this phase, it helps to understand the principles of giving and receiving constructive feedback. The collaboration skills described in this section stress techniques for keeping your team focused on performance so that you do not slip back into a forming or conflict stage of team development. Table 7.2 outlines how the skills and tools discussed in this chapter relate to the various steps of the modeling phase of design. Most have general applicability to all the steps in this phase of your design work.

7.1 PROJECT MANAGEMENT

For most student design projects the main thrust of your work with regard to project management centers around coordinating and clarifying who will be doing what and then ensuring you have the resources and material to get the job done. It is important to realize that you are at a critical point in the life of your project. You still need to move it from concept to reality (or at least provide the details to demonstrate feasibility). Without proper coordination and planning your design efforts can fall short of your expectations. On the other hand, with proper planning you will be in a much better position to meet or even exceed expectations.

7.1.1 Clarifying Role and Responsibilities

It is important to understand that roles developed during earlier phases of design may change as you develop models and prototypes. For example, perhaps one member

TABLE 7.1 Overview of Phase 3: Developing Models and Prototypes

Steps for developing models and prototypes	Skills and tools for formulating solutions			
	Decision making	Project management	Communication	Collaboration
5. Selecting a modeling process 6. Performing design analyses 7. Testing the overall design 8. Revising, refining, and critiquing the design	• Quantitative versus qualitative decisions • Conducting design and critical reviews	• Clarifying roles and responsibilities • Obtaining resources	• Writing progress reports • Providing feedback • Seeking input and feedback	• Managinge role conflict and role ambiguity • Recognizing style differences • Eliminating social loafing

TABLE 7.2 Relationship of Skills and Tools to Each Step in Phase Three of the Design Process—Developing Models and Prototypes

	Phase 3: Design steps			
Skills and tools	**Selecting a modeling process**	**Performing design analyses**	**Testing the overall design**	**Revising, refining, and critiquing**
Project management				
• Clarifying roles and responsibilities	X	X	X	X
• Obtaining resources	X	X	X	X
Communication				
• Writing progress reports	X	X	X	X
• Providing feedback	X	X	X	
• Seeking input and feedback	X	X	X	
Collaboration				
• Managing role conflict and ambiguity	X	X	X	X
• Recognizing style differences	X	X	X	X
• Eliminating social loafing	X	X	X	X

of your group was particularly good at getting everyone to contribute to ideas and emerged as a leader during the brainstorming sessions. That person might not have very strong "tinkering" skills and may need more guidance while the actual pieces of the model are put together. Another person may then emerge as a leader. Working effectively during this phase of design is about understanding who will be doing what, when it will be done, and what will happen if the unexpected occurs. The best place to begin understanding these issues is by taking a look at your Gantt chart or PERT/CPM diagrams (see Section 3.2.1). Review your plans to determine whether they are still realistic and on track. What are the major tasks that need to be completed now that you have formulated your solutions? Have you accurately determined the subtasks and specific steps needed to get the work done? Do your initial time estimates still seem reasonable? Have you prioritized your work properly?

Once you and your team are satisfied with your answers to these questions, it is time to turn your attention toward matching people with the work. There are several issues to consider in relation to this matter. First, what expertise does each task require? You and your team members should determine the skills and abilities it will take to complete each task. For example, the task of getting the materials into the correct state or shape may require skill with a wire cutter or lathe, and the task of acquiring the necessary materials may require the ability to negotiate a fair deal with a supplier.

Another matter to consider is how many people each task requires—like the old joke, how many engineers does it take to screw in a light bulb? Seriously, some tasks may be done more efficiently if more than one person working on them at a time. Others may actually be harder to complete with more people. For instance, is it really possible to test the wheels on a car while driving and examine the alignment at the same time? Consider each task carefully to determine how many people need to be involved for the purposes of efficiency and team cooperation and cohesion. There also may be tasks that can be completed by one person, but the whole team will benefit from being part of the work. For example, it might take only one or two people to test a particular design feature, but changes that are made as a result will affect the whole team, so perhaps everyone should be involved in testing.

Finally, you need to consider two related issues: Which team members have the right mix of skills/abilities to complete each task? Who has the greatest interest in completing each task? It is important to keep in mind that the answers to these questions may not be the same person. You may have a lot of skill in using wire cutters to prepare for wiring, but you may not want to do that during this project. Instead, you might be interested in preparing the presentation materials—a task you have perhaps not attempted before. Remember that this is a growth experience for everyone on the team, and you may all want to try different parts of the project, parts that require skills you still are trying to develop. Remember, too, of course, as new engineering students you may all have to stretch. There may be some tasks for which no team member has any significant knowledge and ability. For these tasks, collaborative work is even more critical; together team members can determine how best to complete the task. As a team, you must balance the interest and abilities of the members with the overriding group goal of a successful design project. Later in this chapter we discuss ways for the group to understand each other's skills, motivations, and work styles, all of which play a role in deciding who should do what.

7.1.2 Obtaining Resources

There are many times in a design process when the design team will have to seek additional resources to continue a project. Here, resources may be related to either time or money. This is why a freshman design team must build some reserves in their budget or why an upper division design team needs to expect the unexpected. In industry, a company may give a team the assignment and resources to take a project through to building a scale model or prototype and doing all the analysis, testing, and refinement so that the company can make a decision about whether to go forward.

In *All Corvettes are Red*, the design team had numerous checkpoints when they had to make presentations in order to have the funding and personnel to go forward. In this multiyear process, the funding required to produce the new Corvette was significant, and periodically management had to make decisions about the project's future. This was done in tough financial times for General Motors, and many of the decisions had far more to do with the state of the corporation's financial health and restructuring than it did with the quality of the work the engineers on the design team were doing.

The team found that they had to seek funding from other organizations within General Motors instead of through the normal Research and Development administrative structure. The Chevrolet division knew how important the Corvette was to bringing customers to showrooms, and they provided funding to keep the project going.

7.2 COMMUNICATION

There are two main communication issues to be dealt with during this phase. The first involves documenting and explaining to others what you have done on the project to date. To help you confront this challenge, we provide you with some guidelines for preparing progress reports that are useful for both team members and others outside the team who have a stake in what your work is and how it is going. Preparing effective progress reports often can be the key to ensuring continued support for your work.

The second major communication issue you face is how to critique ideas and results in a way that points out flaws and yet encourages continued effort. Doing so requires an understanding of how to give and receive constructive feedback. When done poorly, exchange of feedback can lead to breakdowns in communication and even lingering resentment. When done correctly, however, feedback becomes an important part of any problem-solving effort, especially design processes.

7.2.1 Writing Progress Reports

Progress reports take various forms, ranging from a simple "heads-up" to a full-blown external review board (ERB) presentation. In the latter, the project manager or lead engineer is expected to appear before a panel of outside experts with a written report, a flip chart, and slides. The ERB will want to know the following:

1. Is the project on track? They look for proper slack in your schedule and determine whether it can be done on time or not.

2. Is the project on budget? During a formal design review, the experts may suggest changes in hardware or software, adding new capabilities and requirements or new technologies.
3. Is the work being done the best way possible? During a critical design review (see Section 6.3) the experts may suggest better tools and techniques to achieve better performance.

Of course, your best resources are your up-to-date lab journal or notebook, as well as your adjusted design schedule and Gantt chart. A good rule of thumb is not to repeat earlier tasks. Alter them and update them on a chart, but try not to start all over when doing a progress report. Also look to the future—each progress report should anticipate your final report and eventually become part of it.

Before you begin to develop a progress report, consider your audience and the following questions:

1. Who needs the information?
2. When do they need it?
3. What questions or concerns will they have?
4. How will the information be presented?

In a group project, parts of the progress report will likely be drafted and presented by different team members but edited and coordinated by a team leader.

In the commercial world, project managers routinely report to users or clients on a weekly basis. Sometimes the progress report takes the form of a simple narrative, indicating whether the project is slipping, on schedule and budget, or ahead of schedule because of good weather or a technological advance.

Increasingly, the weekly progress report takes the form of a face-to-face or virtual meeting with key team leaders, staff, and users or clients. Such an "all hands on deck" meeting, whether a teleconference or in person, occurs at the same time each week and in the same place or manner.

At times, face-to-face and even virtual meetings (such as an electronic meeting system, CUSeeMe, or VITs) can interrupt the workflow and prove time consuming. A good solution is a brief, perhaps one-page, written progress report from each member to each team leader and one from each team leader to the project manager who, in turn, informs the user, sponsor, or client. The project leader can see at a glance what needs more time and resources.

Engineering professors, too, prefer short, crisp, well-written progress reports, often on a weekly basis. Some professors combine the individual and team reports with a weekly team meeting, perhaps during class time. The weekly progress report is usually graded. At a typical weekly meeting, design students may be expected to demonstrate what they did during the past week. The brief, written report shows, perhaps graphically as well as verbally, what was accomplished. On the basis of that weekly report, anyone else should be able to duplicate what you did and achieve the same results.

Progress Reports for the Assistive Feeding Device Project

In the assistive feeding device project, the students are required to report to the faculty on a weekly basis. The report includes a written report (approximately one page), a formal oral report that is typically delivered by a team of students, and an update of their Gantt chart to show where the problem areas are. The weekly progress reports contain the following items:

- A summary of the accomplishments since the previous week's report.
- A report on the activities of the teams at the other schools. This information is obtained from a weekly video/teleconference.
- An update on any changes in the Web pages at the participating schools.
- A specific identification of any problem areas that were encountered during the previous week.
- A projection of what is to be accomplished during the next week.

Once the students learn the procedure for making the reports, it takes very little time to prepare them, and they provide a professional aspect to the project. The oral reports consist of a presentation of the material in the written report amplified by figures and photographs. The oral reports must be done using MS PowerPoint with a computer projection system. Overhead transparencies are discouraged.

The secret to writing a quality progress report is in keeping a complete notebook. Every step of the design process should be sketched in your notebook: brainstorming, meeting notes, wiring diagrams, hardware sketches, printouts, catalog of materials, trade-offs, your program (turn left, speed up, climb), chassis changes, power train, navigation controls, you name it. The weekly progress report distills these data and presents them in a clear, methodical way. Your revised and updated Gantt chart can be very useful here.

The general structure of a quality progress report follows a simple three-part flow: introduction, body, and conclusion.

The introduction, like the conclusion, is very sharp, preferably just a sentence or two, presenting the main design problem or challenge for that week. Busy professors appreciate writers who get to the point swiftly. A long drawn-out narrative of excuses or apologies is a definite turnoff.

The body of the progress report explains what was done in a way that the effort could be replicated by the professor or an outsider. Precise language is necessary, and the flow should be systematic and methodical. Simple graphs, charts, or diagrams may be inserted to save words and illustrate difficult-to-describe procedures. However, such graphics must be woven into your narrative and not just inserted without explanation, reason, or justification.

The conclusion should be pointed and crisp and should answer three questions: What did you learn? What are the implications or ramifications of your effort? What are your next steps? Keep this section short. Much of what you learned is evident in the body—whether it is evident to you depends on your closing statement.

Style matters. Your prose style here is not literary, nor is it dry and dense. Your report should show your enthusiasm in doing your design work. If it was a dull, mind-numbing experience for you, guess how that progress report will read to your professor.

Aim for clarity in a progress report. Short sentences are better than long ones, although a series of short sentences can be choppy. Aim for balance and flow. Short words are better than long words, but the best, correct word might be long, so use it. Finally, spell-check your progress report and make sure the grammar and punctuation are correct. One glaring error or mistake can give the wrong impression of your work, cause distraction, and trigger a lower grade.

Figures 7.1 and 7.2 provide some tools to help you prepare your own progress reports. Figure 7.1 presents a checklist of topics you will want to address in various sections of a written progress report. You may not have to address all these topics in every progress report, but you will definitely have to address each of them at some point. Notice that the issues listed under each topic are very straight and to the point.

Progress Report Checklist				
Report Topic	Report 1	Report 2	Report 3	Report 4
Cover Page Project Name Who's Doing It Reporting Period				
Opening Paragraph Design Problem(s) Design Solution(s)				
Main Body What Has Been Done What Has Been Considered Team Decisions Sketches				
Design Schedule Ahead Behind Reasons (Gantt Chart)				
Budget Estimated (original) Spent So Far Current Condition (over/under) (Budget Chart)				
Future Work Next Two Weeks				
Spelling, Grammar & Punctuation				

FIGURE 7.1 Design project progress report checklist.

DESIGN PROJECT PROGRESS RECORD		
Report #:	Team Name:	Date:
Design Project Title:		

Activity	Estimated Hours	Actual Hours	Percent Complete 25	50	75	100
SCHEDULING & PLANNING						
Select team leader						
Build WBS						
Prepare Gantt chart						
Make SOW						
Progress report						
DEFINING THE PROBLEM						
Form problem statement						
Identity functional requirements						
List constraints/limits						
Gather library data						
Conduct interviews						
Progress report						
FORMULATING SOLUTIONS						
Brainstorming session						
Evaluate alternatives						
Select potential solutions						
Progress report						
DEVELOPING MODELS, PROTOTYPES						
Prepare scale drawings						
Perform design analysis						
Test preliminary design						
Revise, refine design						
Progress report						
PREPARING AND IMPLEMENTING DESIGN						
Complete drawings						
Graph data for report						
Prepare visuals for oral report						
Type final report						
Prepare oral presentation						
Complete working prototype						

FIGURE 7.2 Design project activity completion record.

If you keep your responses the same way, you will have prepared a progress report that is useful and easy to follow.

Figure 7.2 provides an example of how you might organize a more concise summary of progress. This format can be helpful when all you need to do is provide a very quick status update of what is done and not done without going into detail. The approach can be quite useful as a communication and planning tool within your team. Finally, this format should look somewhat familiar to you. It is essentially a modified version a Gantt chart and our work breakdown structure (see Section 3.2.2).

7.2.2 Providing Feedback

As your team continues to work together, one of the key things members need to do is communicate with each other about the work that is getting done and its quality. Of particular note is the issue of feedback. Feedback is specific information about how the project is progressing. For team members to collaborate effectively, they must be able to provide feedback to other team members in a way that encourages the entire group to grow and work with increasing success. Providing feedback to others is a difficult process, even for the most seasoned engineers. This is because, although much of the feedback you will exchange with your design team members is about technical issues and/or the performance of your design, there is almost always an emotional component to exchanging feedback. Failure to recognize this fact can lead to feedback doing more harm than good.

The first question we have to answer is "What is feedback?" For the purposes of our discussion, feedback is information about work that is being done and how it is being done. Feedback can come in many forms—when you get a grade at the end of the semester, you have feedback; when actors read reviews of their movies, they have feedback; when a manager instructs an employee about how to increase performance on the job, the employee has feedback. Feedback can be quantitative (you are in the top 25 percent of your class according to your grade-point average) or qualitative (the actor is called wooden and disingenuous). Feedback can also be constructive: inspired to help the recipient improve performance by pointing out limitations that can be improved or by specifying the behaviors that made the work successful. Feedback can also be destructive: vague, critical statements about why the person is inadequate, often delivered in insensitive and possibly even threatening tones.

In the case of engineering design all students involved are on an equal footing—no single student is responsible for giving feedback to all others. In fact, all participants should be giving feedback to all others as needed. Some guidelines for giving constructive feedback are summarized in Table 7.3.

Perhaps you have already noticed that members of your group are providing feedback to each other. It is probably worthwhile to consider how well your group is doing and perhaps provide feedback about providing feedback! Think of specific examples for each of these questions, assess how well you and your team members did, and consider techniques that might have improved the situation. Suggestions are given in Table 7.4.

TABLE 7.3 Guidelines for Delivering Constructive Feedback

What	How
Be sure the feedback process is fair.	• Give the recipients a chance to express their view of their work; ask them "What do you think? " or "Would you agree?" • Give feedback to any group member who requires it—do not pick on someone because of a personality conflict. • Use a problem-solving approach—if you are not satisfied with the work someone has done, work together to find a better solution. Ask "How can we fix this?" or "What can the group do to make this better?"
Be sure the feedback is relevant to the project.	• Focus feedback on the work the person has done and not on personality or other things the person cannot change. Try this: "You have been late to our last three meetings, and that disrupts the whole group," instead of: "You are a procrastinator who always waits for the last minute on everything."
Provide the feedback in a timely manner.	• Give feedback as soon as you notice the problem or are encouraged with results. • At the end of the project recap the entire process, but do not revisit issues that have been resolved.
Be respectful of others when giving feedback.	• Remember to treat others the way you would like to be treated. • Provide feedback directly to the person with whom you have an issue and not an intermediary.
Work to understand the constraints that affected the person's ability to complete the work efficiently.	• Ask questions about the situation before casting judgment about the work.

TABLE 7.4 Suggested Methods for Assessing Your Team's Ability to Provide Feedback

1. Do members of your team provide specific examples to show exactly what they like about another person's work?
2. Do members provide both positive feedback and constructive feedback?
3. How long do team members wait after they have analyzed the work before giving feedback to someone?
4. Does your group reinforce good performance as well as discuss areas that require improvement?
5. Do team members work to understand and agree on common goals before providing feedback?
6. Does your group try to understand the situation or external events that could have affected the ability of the person to do the work?

TABLE 7.5 Feedback Your Group May Seek About the Project as a Whole

- Have we identified all possible solutions?
- Were we correct in our assessments of the scientific principles?
- Did we analyze all alternatives without bias?
- Were all group members given the chance to participate equally?
- How effectively did our team listen both to each other and to external sources?
- Did we stick to timelines that we set up?
- Is the model or prototype that we built appropriate? Does it work?

7.2.3 Seeking Input and Feedback

In addition to providing feedback to other team members, you will also receive feedback from each other and from others external to your group. Before you set out to ask "How are we doing?" it might be useful to understand feedback and your possible reactions to it. Perhaps not all members of the group will feel equally comfortable with the process of seeking feedback. Others may be interested in seeking feedback only in order to hear good things about themselves and the group, but they may have no interest in hearing negative information.

First, we explore the types of feedback you and your team members might seek. Team members can seek feedback about their own work or about the group as a whole. Finally, we discuss some ways to seek that feedback in a productive manner. Different types of feedback are summarized in Tables 7.5 and 7.6.

For both the team as a whole and the individual members, the most important thing about seeking feedback is understanding potential reactions to the feedback. A good way to prepare for receiving feedback from others is to assess yourself on those things that you want to receive feedback. Self-assessment is a process where you take a serious look at you own skills, abilities, and contributions. Self-assessment allows you to accept possibly negative information because you have already addressed it yourself. Accepting more feedback is helpful in making changes. Additionally, discrepancies between our self-image and other people's view of us provide us with valuable insight and can help us improve more rapidly, especially when working on a team.

Throughout the sections in this book are mini-quizzes, checklists, and other tools to help you assess your work and skills. Because these tools exist elsewhere, they are not repeated here. Instead, this is a good opportunity to examine ways to achieve better self-assessments. Some guidelines for self-assessment are given in Table 7.7.

TABLE 7.6 Feedback Individual Members Might Seek

- Am I doing the work everyone expects me to be doing?
- Have I contributed ideas in a way consistent with good collaboration?
- Am I helping others when necessary?
- Have I kept to working agreements?
- Have I met the goals the group set out for me?

TABLE 7.7 Suggested Guidelines for Self-Assessment

- Be open to receiving and acknowledging information about yourself.
- Make mental notes of information you get about yourself so you can pay attention to it later.
- Notice if you have a tendency to take credit for success, or whether you blame situations or others for failure.
- Recognize tendencies you might have to ignore negative information and listen only to positive feedback.
- Ask close confidants or friends to help in your self-assessment—they may be able to point out things you never noticed.
- If you think you are doing something that you do not like in others, find opportunities to take particular notice of those behaviors and carefully assess them in your own way of doing things.
- Recognize that, for most of us, hearing negative comments about ourselves is unsettling, regardless of how trivial the comments may be.
- Be familiar with natural reactions to feedback and recognize that you might be: a) refusing to see patterns in your behavior, always having an explanation for your failure; b) resisting the acknowledgment that you have failed or performed poorly; c) belittling the work that needed to be done—"Well, I did not do that part, but it was a very small piece."
- Even if you disagree with the feedback you are getting, accept it as reality for the person who is giving it.
- Use active listening techniques to demonstrate that you have understood what others are telling you.
- Compare your self-assessment to feedback from others and look for discrepancies.
- View feedback as information that can help you solve problems—that means you will have to objectively assess ratings and information provided by others.
- Ask people for additional information and clarification in order to make sure you have a clear understanding of the feedback you have been given.
- Take small steps—accepting even one piece of feedback helps you and allows you to accept more in the future.

7.3 COLLABORATION

By the time your group has arrived at this third design phase, developing models and prototypes, you will have already done quite a bit of collaborating. You have gotten to know each other and have gone through the early phases of group development; you have thought up new ideas while ensuring that all participants had a chance to contribute; you have evaluated these ideas and have made decisions about design solutions. A lot has been accomplished, and as you begin to develop models and prototypes, you will continue to build on what you have already learned.

When your group was in the defining the problem and formulating solutions phases, you were like a task force. You were new to each other and new to the task at hand; in fact, you were probably new to the professor, and might have been new to the college or university when you were working on the project. Because of all this newness, there was a good deal of uncertainty about the project, and you were probably asking questions like "What exactly are we supposed to do? How long should this take? What will I be doing?" and other similar questions. These questions slowly got resolved as your team understood the process and learned to work collaboratively.

However, at this point in the design process, your team is more like a production team—you have a set idea of what needs to be done, what the end result will consist of, and what kind of time frame you are working under. As a team, you will take materials and create a prototype of your design. While you were formulating solutions, you had an opportunity to get to know your teammates, their strengths and weaknesses, and their interests and resources for completing projects. Your familiarity with one another will help you determine who should do what and will help ensure you continue to work together constructively. In this section we discuss some strategies and techniques that will help you build on your collaborative efforts to date. Specifically we will review guidelines for clarifying roles and responsibilities, introduce you to the notion of style differences, and provide you with some tips for overcoming some common team problems like social loafing.

7.3.1 Managing Role Conflict and Role Ambiguity

After clarifying your roles and responsibilities you still need to manage the process of meeting the demands they entail. To do this it can help to be familiar with two common stressors that occur when people work on teams: role conflict and role ambiguity. *Role conflict* refers to the stress caused by having several different obligations at the same time. For instance, if you are like most students you must focus on not just your design project but also your other schoolwork, perhaps intramural sports, and the part-time job you may need. All these activities place demands on your time and can make it difficult to give the ideal amount of time to each endeavor. When working collaboratively, team members must understand the additional demands placed on all the participants. Some tips for reducing role conflict are given in Table 7.8.

Role ambiguity arises when people are not sure about what they are supposed to be doing. As your team assigns tasks to each member, it is important to verify that people are clear about just what they are supposed to do, that they feel capable of doing it, and that they know when they are supposed to get it done. Also recognize

TABLE 7.8 Some Tips for Reducing Role Conflict

- Examine the timeline of the project and compare it with the timeline of other events you are engaged in now.
- Ask to be assigned or choose to work on parts of the design project that fit with your other demands.
- Remember that everyone will be busy at certain times, such as during midterms.
- Redefine your role so that it fits better with other scheduling issues.

TABLE 7.9 Some Tips for Reducing Role Ambiguity

- At the end of each planning meeting, have all persons state what they are expected to do and when they will have it completed.
- Have members work in subteams to ensure that there is enough expertise to complete the task.
- Refer to Gantt charts and other planning tools to ensure that everyone understands timelines.
- Use concrete examples to show participants what their contributions will look like.

the fact that a certain amount of ambiguity is normal, especially when you are in a learning situation. Table 7.9 gives some tips for reducing role ambiguity.

7.3.2 Recognizing Style Differences

The previous two sections dealt with collaboration in terms of managing the work that has to be done during this third phase of the design process. Still, just knowing what needs to be done, who will be doing it, and when it needs to be completed does not ensure a collaborative work environment. You also have to be comfortable managing the day-to-day interaction with one another, or, in other words, managing relationships. One of the keys to managing relationships within a team is understanding that people have different styles or predominant patterns of behavior that define how they approach work and relationships with others.

A variety of style frameworks have been developed over the years, but it is beyond the scope of this book to discuss them in depth. Many of them, however, define a person's style in relation to two polar qualities: assertiveness and responsiveness. Assertiveness refers to how someone goes about expressing his or her own needs. On one end of this continuum are individuals who are likely to express their needs by making demands—"tellers." At the other end are people who are more likely to express their needs by making requests—"askers."

Responsiveness deals with the extent to which people are likely to be cooperative and concerned with meeting the needs of others. On one end of this continuum are people who are highly cooperative or "open." At the other extreme are people who express little interest in others' needs or who can be characterized as "closed."

What Is Your Interpersonal Style? Psychologist William Marston combined these two qualities to create a model of interpersonal style consisting of four dimensions: dominance, influence, steadiness, and conscientiousness. To learn more about your style, take a few minutes to complete the survey in Figure 7.3, which was developed by Richard O'Brien.[1]

If you completed the survey, your predominant style is the one with the highest number in the box. Keep in mind, however, that most of us possess at least some traits from more than one style. Also, remember that this is just a brief sample of one

[1] R. T. O'Brien, "Blood and Black Bile: four style behavior models in training," *Training/HRD*, Jan. 1993.

Step One: Examine the four descriptive adjectives in each of the five boxes below. In each box, rank the adjective that most describes you as "7," the next closest adjective as "5," and the next closest as "3" and the one that least describes you as "1." Base your responses on your behavior when interacting with other members of your design team. When you are done, each box should have four adjectives ranked 7, 5, 3, 1 (no ties).

1. ____ a. stubborn ____ b. persuasive ____ c. gentle ____ d. humble	4. ____ a. determined ____ b. convincing ____ c. good-natured ____ d. cautious
2. ____ a. competitive ____ b. playful ____ c. obliging ____ d. obedient	5. ____ a. assertive ____ b. optimistic ____ c. lenient ____ d. accurate
3. ____ a. adventurous ____ b. life-of-the-party ____ c. moderate ____ d. precise	

Step Two: Transfer your responses to this answer sheet and then total columns A, B, C, and D.

	A	B	C	D
1				
2				
3				
4				
5				
Total				

Step Three: Transfer the totals for each column to the circles within each corresponding quadrant marked A, B, C, and D.

Pros: high-ego strength; strong willed, decisive; efficient, desires change; competitive; independent; practical **A. Dominant** ◯ *Cons:* pushy; impatient; domineering; attacks first; tough, harsh	*Pros:* emotional; enthusiastic; optimistic; ambitious; friendly; talkative; people-oriented; stimulating **B. Influential** ◯ *Cons:* disorganized; undisciplined; manipulative; excitable; reactive; vain
Pros: perfectionist; sensitive; accurate; persistent; serious; needs explanations; orderly; cautious ◯ **D. Conscientious** *Cons:* stuffy; picky; indecisive; moralistic; fears criticism; critical	*Pros:* dependable; agreeable; supportive; contented; calm; amiable; respectful **C. Steadiness** ◯ *Cons:* unsure; pliable; awkward; possessive; conforming; insecure

FIGURE 7.3 Interpersonal style survey.

style inventory. You are likely to encounter many different surveys during your educational and professional careers.

Regardless of what framework you are using, the most important lesson to take away from any interpersonal style instrument is that that there is no single best style. As the pros and cons in each box above imply, different styles have their own strengths and limitations. What is important, then, is to learn how to modify our styles so that we can capitalize on our strengths and minimize the impact of our limitations. An understanding of style differences can also help us to appreciate another person's strengths, even when they are different from our own.

Information about interpersonal style can often help a team gain a better understanding of how and why it works the way it does. This information should then become a springboard for discussion, planning, and adjustment to your team's collaborating. For example, an instrument similar to the one included in this section, was used by a team of engineers that one of the authors worked with. This team was good at coming up with ideas but frequently struggled with implementing them. The instrument revealed that virtually all the team members were either influencers or dominators. A quick review of the pros and cons associated with these styles points out that none of them particularly enjoyed confronting many of the details that go along with implementing a new design idea. The insight provided by the instrument helped them to see their struggle as a team problem, and, instead of getting caught up in a blame game, they developed effective strategies to ensure the details would not fall through the cracks.

What Is Your Work Style? As another example of a style inventory, consider the following brief discussion of differences in work styles. Some people are capable of hours of intense work, perhaps even through the night, without a severe decline in the quality of the work done. Others prefer a slower, more deliberate pace, with adequate time to think through many aspects of the work before proceeding. Still others are extremely productive over a very short period of time, and then they need to break and recharge for some time. Any of these work styles can be effective and, in fact, may result in the same work being completed. However, if two people with different styles are thrust together, there may be problems unless they take the time to recognize their differences and try to accommodate one another's needs. The following table is meant to describe a few styles of work in order to help you understand your work style and perhaps the styles of your team members. Although no exact guidelines exist for putting together people with different work styles, simply understanding that different styles exist can lead to better collaboration.

Think about how you typically do your schoolwork by answering the questions below.

- When you begin a project, is it hours later the next time you look at the clock?
- Are you able to get finished with a large chunk of work before your classmates even get started?
- Are you bored easily by doing the same task repeatedly?

TABLE 7.10 Different Working Styles

Prefers less structured tasks; likes to jump from one idea or task to another	Sample role: teacher/college professor	Sample role: emergency room physician
Focus on same task over and over	Sample role: nuclear reactor control operator	Sample role: air traffic controller
	Slow and steady, can work for several hours straight	Short, intense bursts of productivity, then rest before another burst

- Does it frustrate you when assignments require you to examine many different ideas at once?

And then consider your answers in the context of Table 7.10.

7.3.3 Eliminating Social Loafing

One of the key elements to cooperative and collaborative teams is that they build trust while accomplishing their goals. Trust is built in part through being perceived as reliable and dependable, relative to the tasks you are doing and to your role in the group. When all members of a team trust each other, they are more likely to behave cooperatively. So, trust is truly part of a cooperative cycle: trust begets cooperation and cooperation begets trust.

Occasionally, however, teams are confronted with members who are not following through on their commitments. This tendency for some people to slack off, believing others will make up the difference, is commonly referred to as "social loafing." Most of us have confronted social loafing at one time or another. If you think this phenomenon is creeping into your team's dynamics, it is important that you and your team members address it right away. Remember to use the principles of effective feedback described in Sections 7.2.2 and 7.2.3. Keep your focus on building/restoring a cooperative cycle for your team. Try to avoid letting your emotions get the better of you. To help you confront social loafing constructively, review Table 7.11. It describes some common reasons for social loafing and then provides some suggestions about how to resolve the matter.

CHAPTER REVIEW

Although this third phase of the design process can be highly technical, success still depends heavily on your abilities to manage the overall project, communicate clearly, and work collaboratively. This chapter has provided an overview of some tools and tactics for helping your team stay on track and continuing to evolve into a high-performance unit. In terms of project management, some keys to your success include allocating time, resources, and people effectively. Key communication issues include

TABLE 7.11 Dealing with Social Loafing

Why your team might have a social loafer	What to do about it
Members don't feel responsible for getting the job done.	Encourage participation; make participants accountable for the piece for which they have accepted responsibility. Engage the team member(s) in discussing how they think the work can get done.
Team members do not think their contribution is important to the outcome.	Assign tasks in such a way that everyone has to work on a critical piece of the design.
Individuals don't believe their contribution can be determined.	Make each person and task identifiable; add names and dates to Gantt charts.
Team members don't feel engaged by their tasks; they are not interested.	Assign tasks according to ability, resources, and interest.
Members don't feel obligated to other group members.	Remember the grade at the end of semester.

maintaining good records and being able to provide clear and concise progress reports that keep everyone (both inside and outside of your team) updated on your status. You also need to understand the principles of giving and receiving constructive feedback. After all, critiquing and evaluating are central to this phase of the design process. Good feedback skills will help ensure that information is exchanged clearly and that it is used to encourage improvement, not discourage it. In terms of collaboration skills, understand how differences in interpersonal style and in work style can affect your team and team member communication. Most teams also struggle at times with issues regarding role ambiguity, role conflict, and social loafing. You should be familiar with these concepts and have a general understanding of how to minimize their impact on your team and your design project.

Use the questions below to facilitate your review of the material in this chapter. Also, be sure to complete the behavioral checklist on page 171.

REVIEW QUESTIONS

1. Why does it help to reconsider project roles and responsibilities once you reach the developing models and prototypes phase of the design process?

2. Explain how the processes of testing and modeling and of keeping good records can affect a team's ability to obtain project resources.

3. Explain the precise information needs of those who review your weekly progress reports. How can you go about preparing progress reports that meet those needs?

4. What is meant by the term constructive feedback? Give specific examples from your own team experience that illustrate constructive feedback and also poorly provided feedback.

5. How might differences in interpersonal style impact a team's effectiveness? What are two common dimensions that generally can be used to distinguish style differences among people? How can you use an understanding of style differences to improve the way your team works?

6. Define role ambiguity and role conflict. Provide some specific examples from your team experiences that illustrate each of these concepts.

7. What is social loafing? What should a team do to prevent it from becoming a problem?

DORM ROOM DESIGN PROBLEMS FOR CHAPTER 7: DEVELOPING MODELS AND PROTOTYPES (PROJECT AND PEOPLE SKILLS)

Assignment 18 (communication)

Reflect on your team's work together so far. How do you characterize your team's ability to give and receive feedback? Write a brief summary of your team's effectiveness in this regard. Try to include specific examples that illustrate points raised in Sections 7.2.2 and 7.2.3.

Assignment 19 (project management and communication)

Outline and submit a progress report that updates others on the current status of your design. Do your best to keep it brief and to the point. You may want to use the forms in Figures 7.1 and 7.2 as guides. Include within the report a description of who will be doing what and how you will coordinate your team's efforts.

Assignment 20 (collaboration)

Along with your team members complete the interpersonal style inventory in Section 7.3.2. Use the information it provides to identify ways to improve the way you work as a team. In addition, analyze your team's effectiveness with regard to social loafing. Are there some specific ways your team can improve? If so, make a list of some things you and/or other team members need to do differently.

BEHAVIORAL CHECKLIST FOR PHASE 3: DEVELOPING MODELS AND PROTOTYPES

Instructions: List and then review each behavioral statement in the first column and reflect on your experiences working with your current design team. Then use the scale below to rate your own effectiveness with regard to each behavior. Record your ratings in the second column. Use the third column to rate the effectiveness of your team. You may want to discuss your ratings with the rest of your team. In addition, you can make copies of this form and ask your team members to rate you.[1]

[1] Appendix A contains a development planning form you can use to establish improvement goals in relation to specific behaviors. Focus first on areas with the lowest ratings based on your self-assessment and/or any additional feedback you receive from team members or your instructor.

Rating Scale: 1 = Never 2 = Rarely 3 = Sometimes 4 = Frequently 5 = Always N = Does Not Apply

Decision making	You	Your team
1. Established clear criteria for evaluating design performance		
2. Effectively used design reviews to identify ways of improving the design		
3. Carefully interpreted results from analyses and tests		
4. Was willing to change results or make modifications based on results from analyses		
5. Planned and conducted design analyses in systematic manner		
Project management	**You**	**Your team**
6. Helped identify the right mix of skills and abilities needed to complete work		
7. Clearly defined priorities and work that needed to be completed		
8. Accurately determined what kinds of resources would be needed, including time and people		
9. Followed through on commitments to complete work on time		
10. Helped the team establish and use high-performance standards		
Communication	**You**	**Your team**
11. Developed well-written progress reports		
12. Clearly documented performance results from tests and analyses performed		
13. Solicited feedback and input from others		
14. Provided feedback to others in a constructive and nonthreatening way		
15. Took the time to self-assess performance and improvement opportunities		
Collaboration	**You**	**Your team**
16. Helped to clarify confusion and conflict over roles and responsibilities		
17. Supported others when they needed help or were pressed for time		
18. Recognized and respected individual differences in interpersonal style		
19. Discussed with others how to capitalize on style differences/ similarities within the team		
20. Encouraged accountability among team members and discouraged social loafing		

BIBLIOGRAPHY

LAWBAUGH, W. M., and HOBAN, F. T. *Readings in Systems Engineering*. NASA SP-6102, Washington, DC, 1993.

O'BRIEN, R. T. "Blood and Black Bile: Four Style Behavior Models in Training." *Training/HRD*, Jan. 1993.

MARSTON, W. *Emotions and Normal People*. Harcourt Brace, New York, 1928.

STAY, B. L. *A Guide to Argumentative Writing*. Greenhaven, San Diego, 1996.

CHAPTER 8

PRESENTING AND IMPLEMENTING YOUR DESIGN

You and your team members have modeled, analyzed, tested, and refined your design. Your present design has passed a number of design and critical review sessions. Management (or your professor) has given some early indication that the team's design looks pretty good. You and the other team members have begun to feel a sense of both relief and accomplishment. A rumor is spreading about the possibility of end-of-year bonuses (well, okay, at least a good grade). You've actually started considering how to best celebrate your success.

But the reality is, it's not quite time for the party yet. In fact, there's a considerable amount of work to be finished before the celebration can begin. But before we continue with these last steps, the caution of an earlier chapter is appropriate. Design teams must be both willing and able to revisit any of the steps in any phase of the engineering design process when new information or new perspectives may influence the work in progress. That means we may now want to revisit problem areas and even defects in defining the design problem. As we have said, engineering design has usually been, and will likely continue to be, an iterative process. Keeping in mind where you've been and understanding that you may have to return to an earlier step in the process, you can turn to the tasks of presenting and implementing the design. For many design projects this phase is crucial.

For the typical engineering student, this is the point in the project that requires you to demonstrate that your design works. In many cases student teams have to compete on both an individual and a head-to-head basis. The competitions are followed by an oral presentation and submission of a final report and project notebooks. In the assistive feeding device project, the teams from different schools have to bring their portion of the design to one location, assemble the components, and demonstrate that the device works. This meeting of the teams includes an oral presentation and the submission of a written report. These designs are examples of working prototypes.

In large-scale projects, such as chemical processing plants and civil construction projects, this stage of the design process may be critical design review where the design team makes an oral presentation about what they have designed and submits the project design to the contracting agency for review. In chemical process design, a pilot plant may have been constructed. This could also be true in the environmental treatment of contaminated water where new technology is incorporated.

TABLE 8.1 Overview of Design Phase 4: Presenting and Implementing your Design

Steps for presenting and implementing the design	Skills and tools for presenting and implementing your design			
	Decision making	Project management	Communications	Collaboration
1. Presenting the final design 2. Implementing production 3. Introducing and distributing 4. Following-up	• Dealing with last-minute changes, retrofits, and workarounds • Checklisting • Seeking a fresh perspective	• Ensuring quality management • Applying Continuous Quality Improvement to your design project • Reviewing performance	• Developing presentation skills • Preparing visual displays • Making the presentation • Writing final reports	• Involving all team members • Reviewing team effectiveness • Celebrating success

This phase, like the others, requires carefully executed steps to present and implement your design. Table 8.1 outlines this phase of the design process as well as some key tools and techniques you will want to use. However, in a departure from our approach to describing the previous design phases, we have chosen to use a single chapter to discuss both the steps of this phase and all four related skill areas. Before beginning to read, take a look at Table 8.2 to obtain an overview of how the various tools and techniques best relate to each step in this design phase. As per this table, you will find the description of each of these techniques within the chapter sections explaining the step of the design to which they most relate.

8.1 PRESENTING THE FINAL DESIGN

Presenting the final design certainly is one of the major milestones along the road in the course of an engineering design project. The final design presentation for an engineering team can be compared with the opening night for the ensemble of a new theatrical production—a team of many individuals with a wide variety of creative talents and backgrounds now has the opportunity to show publicly for the first time the results of their combined creative efforts. For many student design project exercises, presenting the final design represents the end of the project or assignment, typically near the end of an academic term. Such presentations, whether given orally or submitted in a formal written report (or both), often constitute the final exam. And even in engineering practice, presenting the final design can bring back to the engineering professional both the memories of and the anxiety caused by final exams in college.

Your design project will be presented to a panel of faculty members or the rest of the engineering design class or a combination of both. You can look at this as a threat or as a challenge; the better choice is to view it as an opportunity to demonstrate your skills and knowledge.

TABLE 8.2 Relationship of Skills and Tools to Each Step in Phase 4 of the Design Process: Presenting and Implementing the Design

	Phase 4 design steps			
Skills and tools	**Presenting the final design**	**Implementing production**	**Introducing and distributing**	**Following up**
Decision making				
▪ Dealing with last-minute changes, retrofits, and workarounds		X	X	
▪ Checklisting		X	X	
▪ Seeking a fresh perspective		X		
Project management				
▪ Ensuring total quality management		X	X	
▪ Planning and practicing	X			
▪ Reviewing performance				X
Communication				
▪ Developing presentation skills	X			
▪ Preparing visual displays	X			
▪ Making the presentation	X			
▪ Writing final reports	X			X
Collaboration				
▪ nvolving all team members	X			X
▪ Reviewing team effectiveness				X
▪ Celebrating success				X

Focus not on yourself but on the design project; better yet, focus on your audience—their needs, interests, and desires. Do they want to hear excuses or lessons learned? Do they want glitz or substance? Do they want to experience the pain and frustration you may have felt, or would they prefer sound logic and clear thinking? Are they more interested in the design problems you encountered or in the solutions you created? The answers should be obvious.

Realize now that your presentation team might not function like an accomplished symphony orchestra or a well-rehearsed stage play production. Invariably, one team member will get sick, lose heart or interest, and not even show up for presentation of the final design. Another might get mad and walk away. Just look back on the collaboration, communication, and conflict management skills you may have mastered by now, and use them.

Look forward, too, in learning about new tools and techniques. The skills discussed below will help you make a polished and professional presentation of your engineering design. Develop and practice these new abilities to reduce your natural fear of getting up in front of others who may be critical of you and your performance. Here is an opportunity to hone your presentation skills. And guess what—you will be using these skills for the rest of your career as an engineer. You may as well learn to appreciate and enjoy exercising them now.

8.1.1 Developing Presentation Skills

According to a recent "factoid" in *USA Today*, people are more afraid of speaking in public than they are of snakes and car crashes. What is it about being called upon to make a presentation that sends shock waves through most of us?

Surely, most of this fear of public speaking is irrational, like fear of flying. For inexplicable reasons, some intelligent people have such fear of airplanes that they knowingly endanger their lives to a greater degree by driving a long distance in order to avoid a commercial airline flight that is much safer.

Overcoming fear of speaking in public is not unlike recovering from the fear of flying. Three main tools are used in the latter: understanding, simulation, and practice. It helps greatly to know your material, do a few dry runs, and then accumulate rich experiences.

Increasingly, final design projects are being presented orally as well as in written reports. Such presentations range from chalk talks to electronic wizardry. Most professors and commercial clients, however, prefer something in between, where the technology does not obscure or overwhelm the message.

Design presentations can be built easily from the lab notebook or portfolio that is amply explained and illustrated. Engineers who are called upon to do a "dog-and-pony show" still rely significantly on overhead transparencies, often called viewgraphs. They consist of both word charts and illustrative diagrams and photographs. The diagrams and photos can come directly from your lab record book, but the word charts have to be created.

Software programs to develop slides of charts, illustrative diagrams, and word charts are readily available to engineering students. Corel makes one called Presentations, comparable to Microsoft's PowerPoint. Both programs provide templates that guide you toward maximum readability in a landscape or upright position, limiting the number of lines and the size of type.

A crowded chart full of small type cannot be read. In fact, such a chart is more of a distraction than a visual aid; so is a chart that is busy, cramped, and set in an odd array of typefaces. Design simplicity is preferred here, charting only the information that is necessary for study or memorization.

Remember that the information needs of the distant reader and a live audience are different. A reader generally wants to be convinced that you have chosen the best possible design solution. A live audience needs to be persuaded and, in most instances, entertained to some degree. Thus, the information might be essentially the same in written and presented forms but the latter is more visual than verbal. People do not "read" an oral presentation like they do a written report, but they do seem to "read" the character and credibility of the public speaker. Ordinarily, how you say it in a presentation is as important or more so than what you say.

The well-wrought oral presentation normally follows a three-part structure of introduction, body, and conclusion like the written report. Unlike the written report, however, repetition is a good practice in speechwriting. In fact, some public speaking coaches advise you to tell them what you're going to tell them, tell them, and then tell them what you've told them. Another approach is to present the design project chronologically. Figure 8.1 shows one team that presents their project step-by-step and into the future.

Visual aids are thus crucial—not so much word charts but rather diagrams and illustrations. If they are in color, then they are all the more impressive. If they are animated or three-dimensional, better still. However, graphics can be overdone. A glitzy but hollow presentation can be visually stimulating but not very persuasive. Consider your audience. Your job is to express, not impress.

(Time Limit – 12 min.)

I. Introduction – (JD) 1 min.
- A) Introduce team members
- B) Give robot name
- C) Tell roles for members in project

II. Ideas and Preliminary Concepts (GF) 4–5 min.
- A) Research
- B) Describe ideas used
  - 1) Chassis, motors, and drive train
  - 2) Scoop and deposit mechanisms
  - 3) Sensors and control including code
- C) Describe ideas not used
- D) Final design

III. Plan for competition (LG) 1–3 min.
- A) Course details – distances and speeds
- B) Avoiding conflicts

IV. Results (MH) 2–4 min.
- A) Individual competition
  - 1) Maximizing points
  - 2) Evaluation – things to change for next round
- B) Head-to-head competition
  - 1) How robot performed
  - 2) Evaluation – things to be changed for production model

V. Summary – (JD) 2–4 min.
- A) The final product – improvements dictated by competition
- B) Plans for production

VI. Questions and Answers – (JD, GF, MH, LG)

Notes: For this presentation, the group will use PowerPoint. There will be 20–25 frames total and these will include text, diagrams, and pictures.

FIGURE 8.1 Oral report outline for actual student presentation.

8.1.2 Preparing Visual Displays

When preparing visual displays, it is important to determine what equipment will be available when you make the presentation. Currently, three types of presentation equipment are commonly used. These are

1. Overhead transparencies or viewgraphs,
2. Thirty-five-mm slides, and
3. Computer-based projection systems.

Overhead Transparencies Overhead transparencies or viewgraphs are the easiest of the media to use, and any room where presentations are made should have an overhead projector available. This is also the lowest tech of the approaches to visual aids. However, it is also the approach that is currently most commonly used today in universities and businesses. Some of the reasons for this are as follows:

1. With the advent of computer software such as PowerPoint and relatively inexpensive color ink-jet printers, it is possible to make high-quality presentations with overhead transparencies.
2. Assuming that a color printer is available, the presentation can be developed and changed with very little lead time.
3. Presentations using viewgraphs are not as vulnerable to equipment failures as presentations that use the other two methods.
4. There is reduced stress on the presenter because he or she is in total control of the presentation media.
5. The presentation can be changed at the last minute by removing viewgraphs. This is sometimes desirable when several speakers are covering similar topics and the one of the previous speakers has already covered some of the same material you plan to present.
6. Viewgraphs are easy to see in a relatively bright room. It is not necessary to turn out the lights when making the presentation.

Although viewgraphs have a number of advantages, they also have disadvantages. One of these is cost. Depending on the printer used, the material (transparency, frame, and ink or toner) for each viewgraph can be costly. Another problem is that the presentation is, by definition, static. It is not possible to incorporate multimedia easily in the presentation unless auxiliary equipment is used. A third problem is that the colors may be dull or washed-out depending on the printer used.

Thirty-Five-mm Slides Thirty-five-mm slides were used extensively in professional presentations several years ago because of the bright colors and the ability to show high-quality color pictures of objects along with words. They are also relatively inexpensive (compared with viewgraphs) to produce in terms of the cost of the media. Although they have some of the same advantages as viewgraphs, 35-mm slides also have some distinct disadvantages. Some of these are:

1. A high-quality camera is required if pictures are to be included in the presentation. For word slides, a copy stand or film recorder is required.
2. Film processing time must be factored into the lead time required for giving the presentation.
3. The order of the slides is not easily changed once the presentation is set up in the slide holder.
4. Slide projectors periodically jam, especially if the slide holders have been damaged.

Therefore, if the equipment for preparing and making the presentation is available, 35-mm slides can be used for a high-quality presentation. If these are not readily available, however, viewgraphs or projection systems are preferred for student presentations.

Computer-Based Projection Systems A few years ago, computer-based projection systems were too expensive for general student presentations. Specially equipped rooms were required that could cost hundreds of thousands of dollars. However, prices are now such that these systems are becoming commonly available in universities and even more commonly available at professional conferences. Projection systems bring to the student most of the benefits of viewgraphs and 35-mm slides and some additional benefits but few of the drawbacks. The initial investment for the projection systems is relatively high; however, as in the case of computers in general, the price of the equipment decreases annually. The newest systems have such high resolution and brightness that they compare favorably with overhead projectors. In addition, there is the advantage that the media cost for the presentation is zero. However, the largest advantage is that true multimedia can be used as part of the presentation. The presentations can include animation effects, film clips, dynamic slide transitions, and programs that run live as part of the presentation. Presentation development systems such as PowerPoint are currently structured to take advantage of the capabilities of projection systems. If one of the team members becomes an "expert" on the use of such programs available, it is possible to prepare a professional presentation in only a little more time than it takes to collect and prepare the material to present.

The major problem with projection systems is that they are relatively high tech and so reliability can be a problem, especially if several people are using (and changing) a common system. It is recommended that a backup copy of the presentation be prepared on viewgraphs just in case there are equipment problems, especially when the presentation is really important. A fundamental premise is to always be prepared with alternative approaches just in case your initial and preferred one does not materialize. Such extra precautions will provide you with additional confidence, will permit you to continue with your plan regardless of defective equipment, and will gain you the respect of your audience.

If a program such as PowerPoint is used to prepare the presentation, the program will accommodate the formats necessary for the different types of output devices. Typically, there are templates readily available for viewgraphs and 35-mm slides. The template for viewgraphs usually works for computer-based projection systems.

Freshman Engineering Honors—Team Project

Visual Aids for Oral Presentations

- The key is Aids—Content organization must come first
- Approach
  - Clearly define Information to present
  - Carefully plan & organize talk
  - Choose visual aids

FIGURE 8.2 Slide using bulleted items.

Each slide or viewgraph typically consists of a frame and contents within the frame. Here frame is used generically to mean the common material that is included on every slide. This is the material included on the slide master when PowerPoint is used. The common material may include school logos, school or project name, framing clip art, general background, and so forth. The framing elements should be simple and should never detract from the contents of slide. Simplicity enables the slide to print faster and requires less toner or ink than when elaborate frames are used.

The content of the slides is the actual material being presented. As indicated above, the contents of the slide should be kept simple. However, the level of simplicity depends on the final purpose of the presentation. If the slides are used for the presentation only, very few words should be included on any individual slide. In fact, some companies have a maximum number of words that they will permit on a given slide, and that number is fairly small. Typically, only key ideas are identified on a given slide, and these are typically identified by bulleted lists. A typical example is shown in Figure 8.2. A simple slide allows the speaker to use the key words as talking points, and the audience tends to listen to the speaker instead of spending a lot of time reading slides.

If the slides are to be printed and distributed to the audience as a record of the work presented or if the slides are distributed to people who did not even attend the meeting, more information must be provided than would be contained in simple key words. This can be done "on" the slides in two ways. One way is to put more information on each slide; that is, instead of key words, use phrases or even sentences. This way is not recommended, however, because the audience will tend to read the slides instead of listening to the speaker. A better way to provide additional information is to include a notes page with each slide. This is a feature in programs such as PowerPoint in which additional information can be written in a special field for each slide. The notes can then be printed with the slides and included in handouts, but the notes are not displayed during the presentation.

In addition to the words on the slides, it is often effective to include graphic elements on some of the slides to emphasize a point. For example, showing a light bulb to highlight ideas is often effective. Programs such as PowerPoint have a catalog of simple graphic elements that can be used to make slides more interesting. However, it is important not to overuse such clip art. For example, if a heavy graphic is used on each slide, it tends to distract from the oral presentation.

When the slides are generated, color should be used to provide interest and for emphasis. However, if the slides are copied as handouts, they will probably be black and white copies. Therefore, the colors should be selected to make black and white copies readable. Some colors print as essentially the same level of gray as other colors in grayscale. This is especially distracting when color backgrounds are used. Such material that may be easy to read on the screen will become essentially illegible when printed in black and white because the background and letters print in very similar levels of grayscale.

Some care should be given to the colors selected for a given presentation. Some colors clash and make the slides unpleasant to view. The manuals with programs such as PowerPoint usually have tips about what colors work well together.

If the topic calls for photographs or animations, use these when practical. For example, instead of explaining an experimental setup in words, show a photograph and explain it verbally. Similarly, instead of explaining how some device moves, show it moving in an animation if a computer-based projection system is used. Depending on the type of animation desired, programs and applets are available to develop the files necessary for the animations.

If an actual experiment is being described, a video clip of the experiment is extremely effective. For example, in the project to develop a robotic arm for children in wheelchairs, the students videotaped the robot performing a designated task. They included a very short segment (less than 30 seconds) of the video for their final presentation of the project. This was far more effective than trying to explain the operation of the robot arm in words or still photographs.

8.1.3 Making the Presentation

After you have prepared the visual aids for your presentation, you need to prepare to present them. A slick set of visual aids does not ensure a good presentation unless the presenter rehearses. There are a number of "do's and don'ts" that can help improve most students' presentation skills. These are listed in Tables 8.3 and 8.4. Most of the do's and don'ts are based on common sense and are self-explanatory, and not all may apply in all situations.

If you prepare a good presentation, the audience should gain information from your talk. In general, do the best you can on the presentation, and be proud of your effort. If you make a mistake or find an error in a slide, simply go on with the presentation. Be upbeat with what you are presenting and do not call undue attention to any problems that may arise.

One last thing to keep in mind is that in most cases your instructor will be evaluating your presentation. If you are like most people this fact will add some additional stress to making a presentation. On the other hand, if you follow the tips and have taken time to practice, you can go into your presentation with confidence. One way to overcome some of the apprehension that goes along with being evaluated is to go through a sort of mock-evaluation while you are practicing your presentation. If possible, ask your instructor for a list of the specific criteria he or she is using

TABLE 8.3 Do's for Making Presentations

Preparation Do's

1. Go over material—i.e., practice before presentation.
2. Use slides that are short and contain key words only.
3. Use word slides to help the flow of your talk.
4. Show some logic in the order chosen for the slides.
5. Use color on the slides for emphasis.
6. Proofread your slides before the talk. Use a spelling checker.
7. Use all horizontal slides. These project best in most rooms.
8. Use the correct pens when preparing overheads if you write on them.
9. Learn how to use the equipment before making the presentation. Turn on the projector and make sure you know how to change slides. If possible get familiar with the room in which you will be presenting.
10. Do not wait until the last minute to prepare slides.
11. Use a laptop computer with a projection system if at all possible; however, have transparencies as a backup if the presentation is really important and no hardware backup is available.
12. Use fonts that are as large as possible. You should be able to print six slides on one side of asingle 8-1/2" × 11" piece of paper and read the text without straining your eyes.
13. Make the fonts on each of the slides approximately the same size to offer continuity to your presentation.

Speaking Do's

1. If you are introduced, thank the moderator.
2. Make a smooth transition between speakers. Often you can comment (positively) on what the previous speaker said.
3. If a moderator does not introduce you, introduce yourself.
4. Specify the main purpose of your presentation.
5. Give a brief overview of the contents of the presentation.
6. Tell in advance when you are about to change topics. (Give the audience some road signs.)
7. Give a brief conclusion to your talk.
8. End by thanking the audience and asking if there are questions.
9. Leave enough time for questions.
10. Before answering a question, repeat it.
11. Specify assumptions.
12. Keep within the time allocated to you.
13. Speak clearly and loudly enough to be heard.

Non-verbal Do's

1. Keep eye contact with the audience.
2. Spend as little time as possible if changing slides (overheads), and talk while changing slides to introduce what the next slide involves.
3. Have a slide on the screen at all times except possibly at the beginning and end of the presentation.
4. When describing hardware, show a figure or sketch of the actual device before showing details.
5. Use a pointer to identify points on the slides. Be careful with it.
6. Point to the items on the wall screen and not on the overhead projector screen.
7. Dress professionally for the presentation.

TABLE 8.4 Don'ts for Making Presentations

Preparation Don'ts

1. Prepare part of your presentation during your presentation—i.e., do not use a blackboard/whiteboard or start with a blank overhead.
2. Use material on which you cannot answer questions. If you do not understand the material, leave it out of the presentation

Speaking Don'ts

1. Use phrases such as "ah" or "ok" to fill in dead spaces.
2. Use terms that are not defined. Either define terms verbally or give the definitions on the slide.
3. Read material directly from the slides. (The slides are there to provide talking points, not as the text for your entire presentation.)

Non-verbal Don'ts

1. Talk to the screen. (Face and talk to the audience.)
2. Stand in front of the screen, thereby blocking the view. (Stand to one side.)
3. Use your hand as a pointer.
4. Use complicated equations if the purpose is to have the audience follow the equations.
5. Point to the transparency with your pointer. (Point to the screen.)
6. Look at your watch. (Put a clock or watch on the table if you need to know the time during your talk.)
7. Switch back to previously shown slides. (Use duplicates if necessary.)
8. Play with the pointer during your presentation.

to evaluate your presentation. Have all your team members use them while you are practicing and discuss your ratings so that you can continue to refine your presentation. Try to get non-team members to critique you as well. If you cannot get specific criteria from your instructor, try using the evaluation form in Figure 8.3.

8.1.4 Involving All Team Members in Your Presentation

A consistent collaboration theme throughout the design process is ensuring that all team members participate. Whether contributing or evaluating ideas or working on the interesting and critical pieces of the project, it is the very hallmark of teamwork that all members get involved. A key reason, worthy of repeating, is that we work in teams to capitalize on additional knowledge, experience, and resources in the name of completing a project.

Often, however, team members do not carry this premise of open participation through to the final end of the project—the presentation of the design project. Sometimes this breakdown occurs because it seems too difficult to give everyone an equal role in the presentation; other times it is because a member has an aversion to public speaking. In either case, it is up to the team to address the issue of everyone participating in the presentation.

There are many reasons why people might resist getting involved in the final presentation of your design. In the following section we explore some of these reasons

Evaluation criteria	Possible points	Earned points	Improvement suggestions
• Introduction of team member	2		
• Statement of purpose of the presentation	5		
• Adequate number of visuals	10		
• Use of visuals (dos and don'ts)	10		
• Quality of visuals (including model)	15		
• Continuity of the presentation	3		
• Participation of team members (perfect score if all participate)	10		
• Clear presentation of recommended design	10		
• Presentation of alternative solutions	2		
• Coverage of budget	10		
• Consideration of human factors	5		
• Effective conclusion	5		
• Poise and professionalism	3		
• Use of alloted time	10		
Overall	100		

FIGURE 8.3 Sample presentation evaluation form.

and discuss things you and your team can do about them. Perhaps the most obvious of these is plain old stage fright—the tendency to feel uncomfortable, even frightened, by the thought of speaking in front of others. As we mentioned earlier in this chapter, more people are afraid of public speaking than they are of snakes and car crashes.

There are several ways to help team members who feel this way. Allow them to talk about their concerns. You may even want to share your own concerns and reassure them that it is natural to feel nervous about public speaking. You can also reassure them by making the following points: 1) If you are students, stress to them that virtually everyone else in the class is in the same position they are and that no one wants to see them fail. Also point out that no one expects perfection. This is, after all, meant to be a learning experience. 2) One of the benefits of team presentations is that no one has to be "up there" alone. If they run into rough spots, there are others who can support them and jump in. This makes team presentations a great way for someone to get acclimated to public speaking before having to go "solo."

There are other more tactical ways to help someone who is gripped by stage fright. For instance, you can assign that team member the most benign part of the presentation, the piece that will surface the fewest questions or is the least complicated.

You can also help by allowing that team member to practice in front of the rest of the team. Doing so helps the partner become accustomed to the idea of presenting in front of others.

Some team members resist getting involved because *they think they do not have sufficient knowledge or understanding of the material to be presented.* There are several ways to deal with this problem. First, team members should recognize that they have a responsibility to help one another learn new information. In addition, hopefully your team's norms are such that people feel comfortable indicating what they know and do not know. When planning your presentation, ask people if there are any topics they would feel uncomfortable presenting. Offer to spend time reviewing material with other team members until they feel comfortable with what they have to present. As with stage fright, it also helps if you reassure team members that others will be there to help out if they run into problems during their portion of the presentation.

Social loafing is a third reason why someone may resist getting involved in the team presentation. As we mentioned in Section 4.2.4, group work sometimes makes it seem easy for people to avoid doing their share because they think other team members will be there to pick up the slack. Social loafing is something that is best confronted directly but in a way that keeps the focus on the problem and not necessarily the person. For instance, instead of calling someone lazy, explain how the person's action (or lack of action) has delayed completion or required others to do extra work. In terms of the team's presentation you should encourage the participant to take responsibility for a specific piece of the presentation. Be sure to indicate that each member will be held accountable for his or her piece. Make it clear that others are depending on them. If they do not follow through with their portion, it will not be presented. It also helps to establish specific deadlines for loafers. Ideally, ask them to be prepared in advance of the actual presentation date so that the team will still have time to make contingency plans. You should also be prepared to impose consequences on social loafers who do not come through. Consequences might even include leaving their name off a presentation or taking more drastic steps like bringing the problem to the attention of your instructor. (If you do have to go this route, make sure you can document the steps you and other team members took to try and correct the problem yourselves.)

Some team members may resist participating because they are *dissatisfied with the outcome*. If you suspect this might be an issue in your team, explore the matter carefully with the individual(s). You do not have to agree with them, but you can use your listening skills to convey that you at least understand their point of view. Whenever appropriate, acknowledge where and when you agree with them. Try to engage them in problem solving by asking them what they want to do to address their concerns at this point in the project. In some cases, participation in the design presentation may be a requirement. If so, you should remind them of this fact. Finally, stress to them that you need their help to make the best of the situation the way it is. Hopefully, they eventually will come to a point of acceptance and join you in presenting your design in the best light possible.

There also may be situations when your team does not want a particular member to participate in the presentation. For instance, if a team member has been a social loafer, the other members may not think that person should have a chance to participate in the finale. If this is the case, your team needs to recognize the spirit of collaboration and encourage the person to participate at long last but indicate that you do not believe the person made a complete contribution to the project. In other cases, your team may be wary of the person's ability to present the project in the best possible way. Again, your team needs to recall the tenets of open participation and allow the person to participate. Just as your team should have assigned roles and responsibilities with growth and new learning opportunities in mind, so should you approach the task of presenting the design projects. As in the case of the stage-frightened team member, the team itself can coach and rehearse to prepare the participant for a successful presentation.

8.1.5 Writing the Final Design Report

In addition to making a presentation, your professor may also ask you for a final report showing what you have learned and how well you learned it. It is estimated that perhaps 10 percent of a government project's total cost is dedicated to documentation. In the typical academic course, a good term paper may mean the difference between an A and a B. In engineering design courses, your professor may ask each of you or each team for a final design report.

A well-written final design report may compensate somewhat for a failed design. In fact, one of the best final reports in a freshman design course recently came from a team leader who was unable to compete in the autonmous robot competition at the end of the course because the team ran out of time. It is apparent that the team learned from its mistakes and will do a better job of scheduling next time (see Figure 5.4).

A good final design report is built from a well thought-out proposal, solid research, complete laboratory record binders or portfolios, and detailed progress reports. These are all topics we have discussed in previous chapters. The point now is to pull them together into a complete and comprehensive document that summarizes what you did, how you did it, and what the outcomes were. Your professor may have specific requirements on format and contents, but here are the typical ingredients of a final design project in terms of three main sections. Many of the specific elements should be things you are familiar with already.

Front Matter

1. Title page, consisting of the title of the design project, course name and number, your name and the professor's name and title, the date, and perhaps the names of the team members.
2. Abstract or executive summary: an abstract is usually just one paragraph and is placed at the bottom of the title page; an executive summary is usually a page-long summary of the results and findings of the project, and placed after the title page.
3. Table of contents, a complete listing of all the material in your final design report by page number for ready reference.
4. List of figures, keyed to each figure's label or caption, also by page number, including photos, diagrams, and illustrations.

5. List of tables, also keyed to headline or description, plus page numbers where each numerical table can be found in the text.

Text Material

1. Problem statement and description.
2. Background, explaining the cause of the problem and the reason for seeking a solution, being as practical as possible.
3. Requirements list, showing key specifications of the design solutions.
4. Constraints put on the design, such as weight, size, budget, time, and environment.
5. Alternatives available to you in terms of materials, equipment, and solutions.
6. Research, including library work, Internet solutions, computer programs, and interviews with experts.
7. Methodology, the procedures and processes used to develop the final design solution.
8. Work breakdown structure (WBS), who did what, when, and why or how.
9. Team dynamics—analysis of how well or badly your design team functioned in meetings, joint work sessions, and remedial efforts like conflict resolution, mediation, and intervention.
10. Test results and analysis of those tests, models, and prototypes.
11. Troubleshooting and remediation efforts.
12. Lessons learned from success as well as failures plus conclusions and recommendations for those who try to replicate your design efforts.

Annex

1. Bibliography, a complete list of library, Internet, and living sources arranged alphabetically. A reader should be able to find all the information readily.
2. Budget sheet, showing all expenditures, especially reimbursibles.
3. Acknowledgments to those who assisted you.
4. Appendix, a place to supplement discussion with material that is illustrative or informative, such as pertinent technical data, correspondence, and research studies.

Of course, not all design professors require all these sections, but they are listed and described here so they can be used as a checklist for completeness. Some engineering design professors may prefer a letter of transmittal instead of a title page, and others may request a full-blown results and interpretation section instead of conclusions and recommendations. Most professors require bound or stapled reports with covers, and a growing number are requesting electronic versions on computer disk, DVD, CD-ROM, or via the Internet. Now that you are familiar with the main components of your final report, take a few minutes to review the example design report in Figure 8.4.

Introduction: The project was to design a robot that could successfully complete an obstacle course while accomplishing various defined goals. The first goal of the project was to have the robot successfully navigate through the course and dodge another robot that was on the course at the same time. The second goal was to pick up various sizes of marbles from a defined location on the course. The third goal was to sort the two different sized marbles, and the final goal was to deposit the marbles in a defined location that would open the exit gate.

All the previously mentioned goals have to be addressed one at a time through a series of design steps. The approach that was taken in this project was to first think of all the possible ideas that could be incorporated to accomplish all the defined goals. Once all these ideas were listed, the group looked back over the list is to see which ideas were feasible and which ideas could be used together in the robot design. A series of preliminary sketches was made with the idea to determine the preliminary design of the robot. The concepts that were deemed acceptable were then built and tested to see if they accomplished their planned objective. The individual components were then put together to complete the robot design. This was tested again and a series of modifications was made from the information that was collected from the testing procedure. The final step was to run the robot against other teams in a double elimination tournament. This sequence of design steps is the format of how this report is divided.

An additional part of the project was to write this report, which documents the steps that were involved in the creation of the robot and why certain decision were made throughout the design process. This report will also show anyone who picks it up how to build the same robot. This is a very important part of any project since, in today's modern workforce, people are changing positions so often and sometimes the projects are given to another individual who has no idea what went into it previously. This holds very true in this specific project since I am not going to be able to complete the work and someone else will need the background on what was accomplished thus far and why certain decisions were made.

Preliminary Concepts and Analysis: This section of the report will discuss various design options that were developed to accomplish the various goals. It will also explain why each component was selected and how it will accomplish a defined task.

Chassis: The first area of the robot that was addressed was the chassis. We wanted the chassis to be as light as possible and as rigid as possible. We decided to use PVC tube for the majority of the chassis design. The PVC allowed us to mount all the components wherever we wanted. It did not limit us like the Erector set with its predefined hole patterns. The PVC at first was thought to reduce the weight, but it did not accomplish this goal since it turned out to be quite heavy. The PVC did, though, have a great advantage in that it was very rigid, and we needed this since the course allowed for interaction between the two robots competing.

The next part of the chassis that was designed showed how the motors were to be mounted on the PVC frame. This section was constructed from Erector set material and bolted to the rest of the framework. We decided to go with a four-wheel drive design since we needed to go over a trough filled with marbles to be picked up. We also looked into using a tractor belt design but decided that it created too much friction and would cause our robot to perform poorly. The preliminary design had a two-wheel drive setup.

FIGURE 8.4 A Sample final report (minus figures) on robotics. *(continued)*

Motor Selection: The next area addressed in the design process was to determine what was going to power our robot. We decided that we needed a motor that would deliver a high amount of torque. We needed torque since the robot was going to be relatively heavy and needed drive up various inclines. We also decided that the motor must be geared internally since we did not have too much room to mount it to the frame, and the motor must be easily mounted to the Erector Set materials. Our group decided to go with servo motors since they fulfilled all our requirements, especially the high torque demand, and they were lightweight and easily mounted to the robot chassis. Before the servo motors were mounted on the frame, they were modified. We modified them by removing the stop notch that would allow the motor to turn only 180° either way. We also removed the potentiometer since we did not use it and rewired the controls. Once this was completed we attached it to the frame.

Gathering Mechanism: The next area that was analyzed was how we were going to pick up the two different sized marbles. After looking at various ideas, our group decided to build a shovel-type mechanism. This seemed the most logical solution since we decided we did not want to use any additional actuators to operate our gather mechanism. Our design incorporated the use of torsion springs that would be mounted to different sections of the shovel. The different sections would allow the shovel to bend. The design was built out of Erector set material. The shovel can be seen in Figure 1 next to the PVC frame. The shovel is lifted up to the separating mechanism by a string that was connected to a servo motor. We decided to go with a servo motor since we needed to lift a significant amount of weight, which resulted in a high torque.

Separating Mechanism: The next area that was addressed was how we were going to separate the two different sizes of marbles. Various methods were used, such as a plate with small holes that allowed the small marbles to fall through while the large ones rolled over and circular pan with small holes around it so the small marbles could fall out. All the ideas that were developed had the same basic concept of some type of small opening for the small marbles to fall out. Our group decided to use a slot that would allow the large marbles to roll over it and have the small marbles fall through the slot. (Figure not shown.) The marbles were then contained in a chute with a gate affixed on the end. Once the robot was in the deposit position, a solenoid valve would be activated and would pull a pin out that held the gates closed. The marbles would then roll out of the two gates and into the designed collection bin.

Navigation: The next area analyzed was how we were going to navigate the course. Our group decided to use three types of sensors to help us accomplish this goal. The first sensors we used were shaft encoders. This type of sensor allowed us to program the distance we needed to travel by counting the number of times a disk with holes passes through the sensing element. These sensors also allow us to make sure the robot was driving straight since a separate motor controlled each side. The shaft encoders and a small program could control the voltage to each motor to make sure each wheel was traveling the same distance. The next sensor we utilized was a bump switch. This type of sensor allowed us to find our position if we got off track or collided with another robot. These switches are mounted on the

FIGURE 8.4 *(continued)*

four corners of the robot. The final type of sensor that was utilized was an IR sensor. This will help us find important locations, such as the collection bin and the depositing bin. The course construction will have IR beams at these defined locations. The robot will be able to sense the IR signal and position itself along the light path.

Preliminary Testing and Refinement: This section discusses the various tests that we performed to check the functionality of each section of the robot design. Once all the test information was compiled, a series of refinement operations were completed and they will also be discussed in this section.

Driving the Robot: The first test performed was to determine how well the robot drove and turned. The results of this test showed that the robot could not turn in a smooth fashion since it was two-wheel drive. These results made us switch our design to a four-wheel design that used chain drive to actuate the front wheels. This design modification worked.

The next test was to see if the robot could run up and down the ramps. We found from the results of the test that the robot was top heavy and the front-wheel assembly was not rigid enough. Moving the handy board to a more central location corrected the top-heavy problem, and adding some cross members strengthened the front-wheel assembly. These design modifications can be seen in Figure 8. Since our group ran out of time, we did not have a chance to install the bump switches, shaft encoders, and the IR sensors to determine the functionality of those components.

Operation of the Gathering Mechanism: This component of the robot design was tested on the actual competition course. We lowered the shovel into the holding bin and drove the robot to the end of the wall. Once we reached the end, the servo motor was activated to lift the shovel out of the bin. The results of this test showed us that this probably was going to be the most difficult task. When we lowered the shovel, marbles got pinned underneath the shovel and would not allow for any marbles to enter. We tried to modify the shovel by shortening the actual length to minimize the area where marbles could have been pinned. We also constructed the shovel out of plastic rather than Erector Set materials to reduce the weight. This did not work either. That is where the robot is now. If our group had more time we would have liked to put some sort of spring-loaded "sweeper" that would push the marbles out of the way so the shovel could be lowered completely.

Separating Mechanism: The separating mechanism was tested by placing a selection of marbles on the top and allowing them to filter through the system. This component worked exactly as it was designed. There is still a very slight chance that a small marble could enter in the large marble chute, but the probability is very slim. The only problem that we encountered during the test was blockage: the pins that held the gates closed got stuck on the wood. We removed the pins and filed them down and also installed a small piece of aluminum on the top of the gate to reduce the friction. There is still a slight problem with the solenoid at times. Our group would have liked to install a larger solenoid, but again we ran out of time. The larger solenoid would have greater strength, which we think would eliminate any chance of the pins getting stuck.

FIGURE 8.4 *(continued)*

- How many of the listed elements does this report contain?
- What, if anything, essential is missing?
- What do you like most about it?
- What do you like least?
- How would you improve it?
- What kind of suggestions might you provide to its authors?

8.2 IMPLEMENTING PRODUCTION

The design has been approved and it's now finalized. The next step is to implement production, which is the process by which the final design is changed from an idea or prototype into a product, process, or structure. In the context of larger design projects, when the design becomes finalized or frozen, changes to the design beyond this point are often more expensive.

Good project management techniques can help you here. Even when your design is frozen, you may run into component failures or deficiencies. Now you can rely on the concurrent engineering techniques you learned about in previous chapters. You surely realize by now that good engineering is an iterative process, not like a cookbook or formula. Even now in Phase 4 you must be willing to deal with last-minute changes in design or requirements. Your client or professor may have changed the rules or may have added or deleted some functional requirement; or you may have discovered some glitch in your design solution.

Implementing production is sort of a climax in the whole engineering process, and it is also a time when unexpected problems and opportunities occur. You must be willing to alter your design solution a bit, if necessary, and go back and revisit earlier design decisions.

All the while you are expected to maintain a balance in design performance, schedule, and cost. Some students (and professionals) will become tempted at this stage to "buy their way out" of a design dilemma. Just throwing money at it rarely solves anything. Stretching out your schedule until you figure out a solution is not practical either at this stage, especially when the semester or quarter is drawing to a close, or, in the commercial world, when financing is time sensitive and clients want to move in or use the product you design.

8.2.1 Dealing with Last-Minute Changes, Retrofits, and Workarounds

If your team has been careful and done a good job of planning and carrying out those plans, your design should be ready for the performance testing or design review. However, even the best of design projects can, and likely will, encounter problems.

You may recall the Mars Climate Orbiter that crashed into the Martian surface in September 1999. This multimillion dollar project serves as one example of a design that failed because a simple last-minute problem went undetected. Specifically, the error was not due to a technical glitch or even inaccurate calculations. The vehicle crashed because the spacecraft design team in Colorado used British units to make

FIGURE 8.5 NASA engineers and technicians preparing the Mars Climate Orbiter for launch. (Photo courtesy of NASA.)

calculations that were programmed into parts of the craft's ground-based navigation software. At the same time in California, the actual ground-based navigation team that was responsible for monitoring the flight was unaware that the craft was navigating based on British distance units. Instead, they were using metric calculations to complete their work.

After successfully completing its long journey to Mars the vehicle was supposed to approach the planet at an altitude of about 150 kilometers (93 miles). This is what the navigation team thought the craft was doing. In reality, however, the altitude was about 60 kilometers (37 miles), which caused it to burn up upon entry. As a senior NASA official later explained, "People sometimes make errors . . . the problem here was not the error, it was the failure of NASA's systems engineering, and the checks and balances in our processes to detect the error. That's why we lost the spacecraft."[1]

As the above statement implies, you need not feel bad if flaws still exist within your design. You must, however, remain thorough and vigilant when it comes to seeking out problems that up until now still may have gone undetected. At this late stage of the design process, these problems normally fall into one of three general categories:

1. Last-minute changes: These are adjustments or changes to the design after it has been frozen, particularly those changes that occur very near the final

[1]D. Isbell, M. Hardin, and J. Underwood, *Mars Climate Orbiter Team Finds Likely Cause of Loss*, NASA News Release 99-113.

performance test or design review. The need for such changes can also occur after a design is in production, and they can be especially problematic when they become necessary just before introduction of a product or opening of a structure.

2. Retrofits: Retrofitting is the process of replacing newly designed components that fail or underperform with previously existing components, which may not provide maximum performance but at least are functional.
3. Workarounds: These are used in overcoming an obstacle of some sort when a retrofit or last-minute adjustment is not possible.

In this section, we try to deal with some things you can anticipate and some things you cannot possibly anticipate.

Early in the design process, you were encouraged to carry along two or three of the best ideas as far as you could so that if something should go wrong, your team would have an alternative approach ready to implement. This is the time when such advice will have paid off. Using the autonomous robot design project as a source for some examples, we now provide some additional definition and insight into last-minute changes, retrofits, and workarounds in the following subsections.

For the autonomous robots, such last-minute changes encompass control components that have to be adjusted because the final test conditions or application conditions have changed. These can include changes in the test conditions (light levels, ambient conditions, spacing, or distances that have changed) or terrain that is different from development conditions or original maps, and changes in internal components (battery charge, for example).

Retrofitting means that a major component does not do what it is supposed to do and an older/newer or larger/smaller component that has more desirable characteristics replaces the current component. Be aware that retrofitting may actually be the focus of a design project as well as a solution for a project in need of a quick fix.

The term workaround can mean many things. In this case we define it as being a process of using what is currently in place but devising a solution for overcoming some sort of obstacle, performance deficiency, or unforeseen requirement.

Fortunately, you already have in your tool kit some of the skills necessary to attack those problems that arise as an urgent need for a last-minute change, retrofit, or workaround. Often, these problems require some new and possibly clever ideas, and these ideas have to be generated quickly. We have previously described the idea-generation processes like brainstorming and lateral thinking (see Section 3.2.1). In Phase 2, brainstorming was used to generate a wide and varied list of design solution alternatives and design parameters. In this phase, the same technique, focused more tightly on the specific last-minute problem, can produce a very usable list of possible solutions. These solutions can then be ranked in order of likelihood of being successful by a decision grid methodology or a force-field analysis. Think about the ways you and your team might generate fresh ideas as we present some examples of last-minute problem situations.

In the autonomous robot design project there are a variety of problems encountered in the preliminary or the final competition. These include the following:

1. Sensors do not work or are torn off.
2. The battery pack charges differently every time.

3. The controller blows a chip.
4. The controller resets because of power demands.
5. The repeatability of the robot in development suddenly changes.
6. Drive mechanism changes.
7. Drive mechanism stops working.
8. Pickup mechanism changes.
9. Pickup mechanism stops working.
10. Sorting mechanism stops working.
11. Course changes.
12. Environment changes (lighting, temperature, humidity).

Numbers 2, 11, and 12 fall under the category of last-minute changes and conceivably are problems that can or should be anticipated. Numbers 1, 3, 7, 9, and 10 fall under the category of retrofitting—replacing a component. Again these are problems that can be anticipated. Numbers 4, 6, and 8 fall under the category of workarounds and these are typically unanticipated problems.

In the autonomous robot design project the following solutions were identified for problems that can or should be anticipated and that would be helpful if they were built into the design from the beginning.

Last-Minute Changes

- Write software with constants defined at the beginning of the program so they can be changed easily.
- Write software with switches so that it can run any portion of the course or any set of portions.
- Make copies of software you are using, particularly the last three of four versions, and have them dated.

Retrofits

- Mount sensors securely but with the ability to remove and replace them easily—try to get chassis to take brunt of impacts instead of sensor or sensor wiring.
- Make the leads for sensors and sensors easily removable.
- Label or color code all leads from sensors, motors, and devices so that a controller can be easily and quickly replaced.
- Design electrical connections and wiring so that the wire conductor does not take the brunt of the load—use shrink wrap, hot glue—and keep wiring inside chassis.
- Make spare subassemblies if affordable—buy spare parts if affordable.

Workarounds

- Design drive train with adjustable belts and adjustable spacing and with design adjustments—cut slots, put in adjustment screws.

- Design drive train to be easily disassembled and assembled.
- Keep things simple.
- Use nonmoving parts where possible—examples are design passive sorter, design fixed pickup if possible.
- Use Loc-tite (or cheap, colored fingernail polish) on threaded fasteners.
- Reduce friction—use lubricants, align axles, leave spaces, design adjustments.
- Let gravity work for you.
- Cut a flat spot on axles, pulleys, and gears if attaching wheels with set screws.
- Have navigation work on one basis—bumping—but backed up by another—time or distance.
- Anticipate competitors' movements.
- Think about interactions—with the course and with other robots.
- Have personnel backups—assembly, programming, adjustments.

Although the preceding list deals with some specific problems relating to small autonomous robots, these can to be generalized for a wide variety of problems. The following list is an attempt to make things more generic.

Anticipated Problems: Electrical/Mechanical

- Make controls flexible and easy to change.
- Clearly label system components—make assembly diagrams.
- Use graphics and written documentation for design—use color or patterns for clarity.
- Make chassis and drive train component parts adjustable.
- Reduce friction.
- Design for ambient cooling.
- Design for "slow but sure," while leaving opportunity to speed up.
- Buy extra components when possible—buy two for every one you need.
- Have alternative strategies for irreplaceable parts.
- Allow time to solve problems and fix parts.
- Make design easy to assemble, disassemble, or change.
- Have your colleagues cross train and work in teams so more that one person can handle different problems.

Anticipated Problems: Civil/Construction

- Allow for inclement weather in design and construction schedule.
- Investigate soil and subsoil as thoroughly as possible—allow time for unanticipated soil problems.
- If soil has to be moved see if cut and fill can be equal—identify alternative sources if soil must be purchased.

- Identify alternative contractors—know track record of current contractor.
- Understand union contracts and work rules.
- Get good legal help for contracts.
- Know and understand codes and restrictions for the project area.

Unanticipated Problems That Require Workarounds

- Avoid them if you can.
- Make a list of all things that can go wrong by brainstorming about all the things that might go wrong and have a fix or two for every item on the list.
- Know your team members and how to work on unanticipated problems with processes for making quick decisions.

An unanticipated problem such as the following could be solved by a work around. The wheelchair on which the robot arm is to be mounted may be delivered with different dimensions than the test unit. The attachment may be different and require spacers so that the original mounting bracket will fit, or something like adjustable hose clamps may have to be fastened to the chair.

In designing one of the lab experiments for Ohio State freshmen, the designers chose a disposable camera and found that the manufacturer had changed the internal circuitry. This required rewriting the lab exercise as well as new explanations for the lab procedures.

8.2.2 Checklisting

Pilots do it. NASA engineers do it. Busy, organized people do it. Even that fellow who lives at the North Pole does it. They make a list and check it off. Pilots run through a checklist before takeoff or landing to be sure all necessary actions have been taken to ensure a safe departure or arrival. In a similar way, the countdown to a space shuttle launch is an automated checklist in which literally thousands of actions are made or verified under computer control before liftoff. The daily schedule of busy people is often organized around a list of things to do.

Making and using a checklist has a wide range of application in the engineering design process. It is a particularly useful skill in both decision making and project management. Did you notice we have actually been using checklists from the beginning of our discussion of the engineering design process? Each chapter dealing with a phase of the process has closed with a behavioral checklist. Many of the tables in this and previous chapters are a kind of checklist. And if you glance back at the previous section on dealing with last-minute changes, retrofits, and workarounds, you will see that we have used a kind of checklist. We will further develop the checklisting skill in the context of generating ideas for last-minute problems that may occur just before or during implementation of a design.

Checklisting as a strategy for generating ideas, or creativity, was developed by A. F. Osborne whom we recall also introduced the idea of brainstorming. As he describes it, checklisting is a procedure for using words and questions to trigger creative thought. Usually, these triggers focus on possible changes to an existing product,

TABLE 8.5 Checklisting Trigger-Word Categories and Examples

Categories	Examples
Change quantity	Increase, reduce, lengthen, shorten, deepen, combine, fasten, assemble, fractionate
Change order	Reverse, stratify, sequence, arrange, alternate
Change time	Quicken, lengthen, slow, endure, renew, synchronize
Change state	Harden, soften, straighten, curve roughen, smooth, heat, cool, solidify, liquefy, vaporize, pulverize, lubricate, moisten, dry
Change relative position/motion	Attract, repel, lower, rotate, translate, oscillate, raise

process, or system that may have been produced in a brainstorming session. We present an example from G. Voland[1] that has direct application to the last-minute variety of problems. Remember you are seeking a solution to a problem presented by the urgent need for a last-minute change, retrofit, or workaround. And while the range of possible problems is very wide, often they are focused around changes that involve *quantity* (increase or reduce), *order* (reverse, stratify), *time* (quicken, lengthen, synchronize), *state or condition* (harden, straighten), or the *relative motion or position of components* (attract, lower). Table 8.5 summarizes the trigger-word categories and provides further examples of trigger words. Table 8.6 provides a list of checklist trigger questions that can be used to attack a last-minute problem.

8.2.3 Seeking a Fresh Perspective

Still struggling for a solution? You might want to try seeking input from someone who is not involved with your team's design project effort. Carefully explain the problem to this person and see if some new insight, idea, or understanding can be obtained from a fresh perspective. It sounds simple, and it is. But often it can provide that one idea or insight of which neither you nor your team members were able to imagine.

There are many examples of how outsiders or newcomers, unrestrained by existing patterns and habits, provided a fresh perspective that led to new solutions. Consider this story described by Joel Barker in his book *Paradigms: The Business of Discovering the Future*:

> *In the 1930's, the General Electric Company supposedly had a practical joke that it played on every new engineer in its incandescent lighting group. It went like this: Each new engineer began his job by meeting with the director of the division. The director would turn on an incandescent light bulb and observe, "Do you see the hot spot in this bulb?" (At that time, you could see the filament in bulbs even though they had a coating on them.) "Your job is to develop a new coating that smooths the illumination out so that the entire surface of the bulb glows in a uniform manner."*

[1]G. Voland, *Engineering by Design*, Addison-Wesley, Reading, MA, 1999.

TABLE 8.6 Checklisting Trigger Questions

What is wrong with it?	What does it fail to do?
What is similar to it?	In what way is it not efficient?
Why is it necessary?	How can the cost be reduced?
What can be eliminated?	Can it be misused?
What materials can be used or substituted?	Under what conditions must it operate?
How can it be assembled more easily?	What is the expected or required reliability?
What, if anything, can be eliminated?	Who will use it or operate it?
Is it safe?	How can it be improved?

The new engineer would then go off to tackle the problem, unaware of the fact that everyone else knew the problem could not be solved. After several weeks of struggling, the engineer would admit defeat and then to hearty laughter of his colleagues be told that the task could not be done.

It was a good joke . . . until 1952 . . . when a newly hired engineer returned to the director, screwed his bulb into the socket, turned it on and asked, "Is this what you were looking for, sir?" And, as the director looked at the first bulb that met his impossible conditions, he reportedly said, "Ah, yup. That's it."

For student design projects, your design professor and other professors or students at your school can provide the needed fresh perspective. And you may be pleasantly surprised. The process of carefully explaining the problem to another can help you understand the details so that you arrive at your own new idea.

8.3 INTRODUCING AND DISTRIBUTING

In industrial and commercial settings this step is the point in the design process when production is underway. Those members of the team who are primarily responsible for introducing the design to the marketplace will have become very busy with the many tasks that must be completed for a product to be successful. In the case of the design of a process or system, this step in the engineering design process may have some reduced emphasis. Even so, there are processes that produce a product (like refining crude oil to produce gasoline) and systems that provide services that can be marketed in ways similar to a product (like building a new mass transit system to provide transportation services to the general public).

Here is where profits are generated and payrolls are met. Products of even the finest in engineering design have to be marketed skillfully. In fact, a number of engineering and technology organizations use completion of this step in the design process as metric of a design team's overall performance. For instance, one criterion that Hewlett-Packard uses to evaluate and reward design teams is time to market. They define this criterion as the amount of time that elapses from when a design project starts (problem definition) to when it is released from manufacturing (the point at which it begins being distributed in the market).

Traditionally engineering design students rarely have had to worry themselves with this part of the process, but it is, of course, crucial in commerce and industry. Therefore, you should try to develop a general understanding of what this transitional step entails. Because of the importance of this step, your professor may even have asked you to outline some of the implementation issues associated with your particular design project. Even if he or she has not asked you to do so, we believe it worth your while at least to give this topic some thought.

An interesting phenomenon is occurring in today's marketplace. Engineering professionals are not off the hook or out of the picture entirely once they hand off their drawings to the manufacturing or construction team. They are increasingly involved in production as part of a larger team, even if only to troubleshoot or advise them. In addition, they are called upon more and more to help with introducing and distributing the product or process they design.

You can see design engineers showing up at open houses or demonstration sites to explain such things as the total quality management or continuous quality improvement (see Section 8.3.1) that went into their design of the new automobile or computer system. Using all the communication skills they can muster, these engineers are now called upon to answer queries from media representatives, potential buyers, and clients whenever a new design is introduced.

In the future, you will see even more design engineers who make formal presentations to marketing groups, advertising agencies, public interest groups, civic organizations, and engineering societies about their work. After all, who knows more about the product, process, or system than its design engineers if they have followed the design through all its stages and coupled with manufacturing? Your oral presentation (discussed in Section 8.1) is good practice for this more outward profession. An important element is to recognize the degree of technical versus general detail that may interest your particular audience.

8.3.1 Ensuring Quality Management

Quality management is a key aspect of ensuring that design introduction and distribution stay on track. Over the years, our concepts of quality have changed. In years past quality was considered the job of the blue-collar worker on the shop floor, and quality was inspected by the white-collar supervisors. Flaws in design, defects in manufacturing, and substandard quality were blamed on shop-floor workers who as result may have been inclined to either make excuses or even hide defects in order to avoid negative repercussions. This system did not produce great quality.

Instead of inspecting quality at the end of the entire process or project, today's project managers tend to build quality into the process from the very start through a team approach. Quality becomes everyone's business, and engineers as well as shop-floor personnel are given credit, not blame, for exposing problems early for corrective action and improved quality.

Today, both life-cycle cost and the customer drive quality. Ultimately, quality is now determined by the customer, not by the worker and certainly not by the all-knowing manager, it is based on several useful factors:

1. Economics: The product is readily available at a reasonable cost.
2. Ergonomics: The product is user-friendly, attractive, and suitable to the senses (sound, feel, and smell).
3. Operability: The product operates safely and efficiently.
4. Reliability: The product performs without failure; it needs few if any repairs over its life cycle.
5. Dependability: The product performs once (if disposable or intended for a single use before being recycled) or repeatedly at the same level of quality for a specified period of time and frequency.
6. Maintainability: The product is easy and cheap to maintain over its expected lifetime.
7. Disposability: The product can be refurbished or recycled and is friendly to the environment.

Total Quality Management (the buzzword in the 1980s) thus is considered not at the end of your engineering design project but at the very beginning. You should be looking for quality in the team members you associate with, the processes you choose, the materials you select, the alternatives you evaluate, the decisions you make, and the knowledge and skills you develop.

However, just as there are grades for student projects, there are gradations in quality. Zero-defect top quality may be impossible to achieve on a limited budget. You can aim to just get by with marginal quality, or you can aim for higher quality assurance. One important point of TQM is the M (management): Quality needs to be managed. It does not just happen—it is deliberate, planned, and balanced with other factors such as risk and cost.

A useful analogy of this notion of balance may be found in a chocolate chip cookie recipe. Too many chips make the cookies gooey. You could use butter instead of margarine to enhance the quality but you also add cholesterol, fat, and calories to the cookies. The perfect recipe calls for a balance of flour, eggs, sugar, milk, butter, chocolate chips, and nuts. Too much or too little of any one ingredient throws off the recipe. Similarly, an engineering design project typically calls for balancing competing needs. A project that is completely failsafe is either too expensive to build or, in the case of a student project, too long to fit into a semester or quarter. Accommodations have to be made. Tradeoffs need to be considered in the mix of ingredients.

Hardly anyone talks about TQM anymore, it should be noted. Now the talk is on Continuous Quality Improvement, taking those small steps in a team environment for incremental improvement. CQI is more than tweaking, which is usually fine adjustments made to the final design product. Instead, CQI is directed at the process as much as at the product.

CQI advocates claim that they can improve the quality of the final product by making continuous improvements in teamwork, the work environment, the structure and organization of the work, decision-making processes, and customer relations. They view CQI as a spiral rather than a linear process. They are not afraid to go back a few steps to resolve a conflict, rework a wiring diagram, or reconsider an earlier decision on torque or power source.

Another variation of TQM and CQI is a process known as reengineering. It calls for major, rapid changes. Thus, instead of gradual, continual change or modifications for improvement, we experience radical design changes, abrupt change of direction, and massive staff realignment—a whole new way of doing business (a paradigm shift). Critics of corporate reengineering claim that the changes were often too many and came too soon, causing disorientation and a demoralized spirit among those left behind with additional work and responsibilities. In many instances, the bottom line did not improve as expected.

Still another variation is the ISO 9000 series. The International Standardization Organization, based in Geneva, Switzerland, has established a continuous cycle of planning, program control, and documentation for quality assurance of products and services. Companies can become certified in ISO 9000 when their management systems demonstrate the right kind of planning (setting achievable goals and objectives, including lines of authority and responsibility), control (steps and procedures to meet the goals and objectives, including workarounds or retrofitting), and documentation (analysis and reporting on goals and objectives). Like CQI, ISO 9000 focuses on quality process and not just the final product or service.

8.3.2 Applying CQI to Your Design Project

Engineering design students can use CQI to improve their processes continuously by following the procedures described in this book. Experience will show that too little time, attention, and resources are spent in Phase 1, defining the problem. CQI suggests an interactive process, continuously reevaluating how flaws in the problem statement can cause problems later on in the design process. Research may have been too sparse, working agreements may not have been clear to all team members, and laboratory record books may not have been set up or maintained adequately.

Formulating solutions, the second phase, also should be revisited at later stages in the design process. It may be that not enough alternatives were considered before a potential solution was selected. Gantt charts and PERT diagrams may have been constructed but not maintained properly, allowing the design project to drift. More than likely, team conflicts may have been set aside, rather than confronted and dealt with, resulting in blame or accusations for design flaws. CQI can catch those problems before they damage or destroy the team and the project, if team members are open and aggressive about quality.

When developing the models and prototypes of the third phase, were they adequate for sound, quantitative analysis, or were they rigged or otherwise not valid? When roles and responsibilities are continuously clarified, everyone knows what is expected from everyone else and when. CQI demands that we reflect on our team's total performance and improve it when necessary.

Even upon presenting and implementing the design, we need to evaluate the efforts continuously. The laboratory record book is a good place to do this. Jot down the lessons learned. Document the workarounds, successes, and even the failures to show that you understand the process of engineering design.

The final report is not the place to blame others for your failure or make excuses. Instead, you can show what you did and what you could have done or should have done in each phase of the design project:

1. Defining the problem
2. Formulating solutions
3. Developing models and prototypes
4. Presenting and implementing the design

For each step of the process you can analyze it in terms of the four professional skill areas we discuss: decision making, project management, communication, and collaboration. If you are human, you will invariably find some problems in that last skill area, but the point here is not to dwell on the problems alone. Focus instead on solutions or possible solutions, and analyze them fully.

This notion of planning and practicing your own performance should be scheduled in your Gantt chart or planning document in the first phase of the design project. Your efforts at this point can easily feed into your team's presentation and implementation of design.

8.4 FOLLOWING UP

The step of following up includes a variety of activities that generally fall under the category of services available to the customer for the life of the product. These services may include installation, training, maintenance, and repair of a product or structure for the consumer.

It is at this ending point in the engineering design process that the ideas or the motivation for a new and improved product or process logically lead back to the beginning point of the engineering design process. Feedback from customers on their experiences with the current product often lead to development of improved or even completely new products.

The smart engineer and student learn from their team's failures and successes. Although dissatisfied clients and customers may be unforgiving, engineering design professors often have a different view of failure—as a learning device.

If you are able to detect and understand the serious flaw in your less-than-successful design, you have learned a valuable lesson that will benefit you later. Learn those lessons now, in design class, before it is too late. NASA engineers, for example, dread the notorious failure review board after a mission accident. They know the board will point the finger at someone along any stage of the engineering or manufacturing process. The designers of the mirrors that were not fully within specifications or tolerances when deployed for the original Hubble space telescope were eventually identified. The National Transportation Safety Board routinely spends millions of dollars and work hours tracking down design flaws in aircraft and automobiles. The Consumer Product Safety Commission finds dangerously designed toys, appliances, and tools and orders corrective actions, including taking them off the store shelves. The Food and Drug Administration does the same for food and pharmaceuticals, and the Occupational Safety and Health Administration cites badly designed workplaces and projects.

Just as you can expect to go back and redesign a flawed product or process, you can go back now in Phase 4 and learn from your errors in software, hardware, and calculations. The wise student exposes these errors in some kind of final report or design project evaluation and then goes two steps further—finds the solution to his or her design problem and documents it.

8.4.1 Reviewing Performance

In other words, your design project is completed, but the learning is not. Certain project management evaluation tools and techniques help you get the most out of your experience and prepare you for your next one. An older but simple model of project management consisted of four parts:

1. Project planning and scheduling
2. Project design and engineering
3. Project implementation
4. Project evaluation.

Previous discussions of project management skills in this in this text focused on the first three of these areas. Now it is time to discuss the fourth area. It is important to realize that project evaluation is not just about judging whether something is right or wrong. A more important role for project evaluation can be defined in terms of learning and planning for future efforts. A common term used by many companies today is *organization learning*. A simple definition of organization learning is "systematically figuring out what works and/or what works better." The main idea behind this concept is that, for a company to gain and sustain competitive advantage, it must develop ways to learn and apply new knowledge faster than its competition does. Implicit in this learning process is ongoing evaluation and review.

In an academic setting, evaluation becomes all the more important for teaching and learning. In an engineering design course, your professor evaluates your performance with a letter grade and possibly a job recommendation. However, that does not prevent you from reviewing your team's performance and your own, whether your professor requires it or not. In the real world, you advance or stagnate on the basis of performance reviews. In the case of particularly bad reviews, you may be looking for a new job or a new line of work. In government and industry, performance reviews are treated very seriously.

Performance reviews can be confidential (formative) or official (summative) or a combination of the two so that workers can learn and grow from the evaluation and use it for promotion, a pay hike, bonus (merit) pay, or a commission.

Performance evaluations run the gamut from weighted checklists to narratives written by supervisors to top management. It is pointless to describe all the mutations here, but we can suggest that you use this textbook to create your own personal performance review. You can give it to your professor in the form of a written final report, or you can use it to learn from your mistakes, especially if you have the same professor next year. In any review of your own performance, two reminders may be helpful. First, try to *measure* your performance by counting things such as hours, money, or points earned. All your goals, set in the initial planning phase, should be measurable to begin with. Second, be philosophical about your performance. Instead of making excuses or blaming others for shortfalls in your design project, look at each design activity as an opportunity to learn, whether you succeed or fail in a task. Look at the bigger picture and see the engineering design project as a closed system where everything is interdependent.

8.4.2 Reviewing Team Process Effectiveness

In addition to reflecting on the overall quality and effectiveness of your work output, you need to reflect on the quality of team member interactions and overall team dynamics. In this section, we focus specifically on how to evaluate collaboration skills—yours, your team members, and your team overall. This includes summarizing some of the key things you learned about collaboration, identifying what went well, and also determining ways you might improve in the future.

This kind of review may seem like an unnecessary step. After all, upon submitting your report and/or making your presentation, your work is done. Technically, this may be true, but you will be missing out on a valuable opportunity. If your project has been like most design projects, there has been little time for reflection. Instead you have been dealing with a series of immediate issues and problems just to get the work done. In this regard, your experience was not all that different from what professional design teams in industry contend with on a daily basis. Because time is always a valuable commodity, most successful team-based organizations have established systematic procedures to help their teams evaluate process and performance. By doing so they enable teams and team members to better understand their strengths and improvement areas. This type of review, in turn, increases the likelihood that team efforts will be even more successful and efficient in the future.

There are several ways you and your team members should go about evaluating your work together. Hopefully you have been using the behavioral checklists at the end of each chapter in this book. Discuss your ratings with your fellow team members. Try to identify any trends or patterns. For instance, perhaps there was one skill area that received consistently low or high ratings across all four phases of your design work. You also should review your team's performance in relation to your team's working agreement (see Section 2.2.2). Ask yourselves the following kinds of questions: Which of our ground rules did we adhere to? Which did we have the most difficulty living up to? Which were most helpful? Why? Which were least helpful? Why?

Another useful question to consider when assessing your team's performance is to ask yourself, "Would I want to work with this same group of people again? Why or why not?" When considering this question, however, be careful not to get into a blame game. Remember the principles of constructive feedback we discussed in Chapter 7. Try to characterize your reactions to this question in specific and behavioral terms. Avoid making broad generalizations, either good ones or bad ones.

In addition to considering how you felt about others, you are likely to learn the most by reflecting upon your own behavior and experiences. Try to answer questions such as the following:

- What did I learn about myself in terms of working on a team?
- Based on this experience what do I like most about team projects?
- Based on this experience, what do I like least about team projects?

- How was this experience similar to or different from other team activities which I have participated?
- What are some ways I was particularly helpful to this team?
- What are some ways I could have been more helpful?
- What are some ways I might make working on a team an even better experience for me in the future?

Ideally, you and your team should set aside time to have a debrief discussion about the team's performance. In addition to considering some of the specific questions, a more basic approach is to ask team members to think about and complete the following statements:

- The biggest challenges this team faced were ...
- Some ways our team was highly effective are ...
- Some ways this team could have been more effective are ...

Responses to these statements can be shared openly or, if necessary, anonymously. In any event, make certain your debrief discussions are focused on growth and learning and not on venting for your own sake. Debrief meetings are also a good time to exchange feedback to one another. After all, team members are usually in the best position to observe each other's strengths and improvement opportunities. To structure your discussion consider modifying the preceding questions so they focus on individuals instead of the team.

8.4.3 Celebrating Successes

There is one last collaboration activity left for you and your team members. Form a circle with your fellow team members. There should be slightly less than an arm's length between you and the people on either side of you. Turn clockwise so that you are facing the back of the person to your right. Now extend your arm (either one) and place it lightly on the shoulder blade of that person to the right. Once everyone's arm is extended, pat one another on the back in recognition of a job well done!

Seriously, acknowledging your own and your team's accomplishments is important. You have been through a lot together. Hopefully you have learned a great deal and grown professionally and personally. As we mentioned in the previous section, your team (like most) probably find improvement areas but this critique should not stop you from taking time out to acknowledge your accomplishments. You deserve to take some pride in what you have done. In fact, doing so often helps to energize team members before they move on to greater challenges, either together or as members of new teams. One way to focus on your team's success is to write your team's resume. What would you say or write about your team to show off its accomplishments? End your work together by asking team members to prepare their own version of the team resume. Then sit back and enjoy the accolades.

CHAPTER REVIEW

In this final chapter we have presented the basic tools and techniques for presenting and implementing your design project. You have discovered that the work is not quite complete until you have made a group presentation and compiled some kind of final report.

Phase 4 of the engineering design process is of particular interest to students because their final grade hinges on a polished oral presentation and a complete final report. The design may have been flawed, but learning is paramount here. What are your lessons learned? Can you find adequate solutions to the problems you may have encountered? Evaluation and assessment are hot items today in the academy and in the professions. This chapter is ideal for self-assessment as well as performance measurement.

Many engineering design courses end with a gala competition and judging. Good luck, but do not forget to say thanks to your team members and maybe even your professor if you have truly advanced your skills and knowledge.

To review this chapter, answer the questions below. In addition, review the behavioral skills checklist at the end of the chapter and use it to assess how well you are applying the skills and tools discussed in this chapter. Next, use the forms in Appendix A to help you summarize and reflect on your overall behavioral effectiveness throughout the entire design process.

REVIEW QUESTIONS

1. What are the four steps in Phase 4, presenting and implementing the design?

2. Explain the relationship of skills and tools to each step in Phase 4 of the design process.

3. Decision making at this stage ordinarily involves two distinct ways to deal with last-minute changes and problems. What are they?

4. Distinguish between TQM and CQI. Which is preferable in today's work climate?

5. Outline a typical oral presentation. How can you develop the skills necessary for a scintillating presentation?

6. What are the advantages and disadvantages of the three main types of visual display?

7. Outline a typical final design report. What ingredients are necessary and optional in front matter, main text, and annex?

8. What are some ways you might apply collaboration skills during this final phase of the design process?

DORM ROOM DESIGN PROBLEMS FOR CHAPTER 8: PRESENTING AND IMPLEMENTING THE DESIGN

(Decision making)

Prepare your dorm room design model for one final review. Make a list of any issues/aspects of your design that potentially could pose last-minute problems. For each of these areas, try posing the checklist trigger questions listed in Table 5.4. Conclude by articulating key contingency plans with which you and your team should be familiar in order to work around any last-minute problems.

(*Communication and collaboration*)

Construct the presentation you will make to present your design. Assume you are presenting your concepts and recommendations to the director of student housing and want to convince her to adopt your model as the standard design for all dorm rooms on campus. Plan on having between 30 and 40 minutes to make your presentation. As a team, decide what visual display system(s) you will use; where, when, and how to use graphics; the specific things you will say about your design; and the roles of team members in the presentation.

(*Project management and collaboration*)

Begin the process of evaluating the key outcomes from this project. In terms of overall performance, describe in your own words what you think are the major lessons learned about design in general and also specifically about designing a dorm room. Among other things, consider your team's ability to plan and stay on schedule. Also try to determine the extent to which your final design turned out the way you expected. Indicate what you think were pivotal turning points affecting the success or failure of your design. Prepare a set of recommendations that you would provide to others in order to help them complete this same project effectively. Use the material in Appendix A to summarize some of your team's main strengths and areas for improvement.

(*Communication*)

Combine your project evaluation materials with the other records you have been keeping throughout the project in order to create a final design project report. Make sure your report covers the topics listed on pages 187 and 188 of this chapter. As you review your report ask yourself questions such as:

- Have we provided enough information so that others could repeat the process and steps we took?
- Have we conveyed a clear rationale for the decisions we made and alternatives we selected?
- Have we clearly specified what we have learned from this experience and what we might do or learn in order to be more effective?

BEHAVIORAL CHECKLIST FOR PHASE 4—PRESENTING AND IMPLEMENTING THE DESIGN

Instructions: Review each behavioral statement in the first column and reflect on your experiences working with your current design team. Then use the scale below to rate your own effectiveness with regard to each behavior. Record your ratings in the second column. Use the third column to rate the effectiveness of your team. You may want to discuss your ratings with the rest of your team. In addition, you can make copies of this form and ask your team members to rate you as well.[1]

[1] Appendix A contains a development planning form you can use to establish improvement goals in relation to specific behaviors. Focus first on areas with the lowest ratings based on your self-assessment and/or any additional feedback you receive from team members or your instructor.

Rating Scale: 1 = Never 2 = Rarely 3 = Sometimes 4 = Frequently 5 = Always
N = Does Not Apply

Decision making	You	Your team
1. Attempted to anticipate potential problems in advance		
2. Developed and used a checklist to help ensure the readiness of your design		
3. Sought input from others outside of your team in order to enhance overall decision quality		
4. Remained calm in the face of unexpected results or problems		
5. Used brainstorming or other idea-generation techniques to solve unexpected problems		
Project management	**You**	**Your team**
6. Regularly reviewed team and individual performance throughout the project		
7. Can articulate lessons learned from this project, both good and bad		
8. Can specify ways to improve performance and results in the future		
9. Kept good records and documented work performed during all phases of the project		
10. Helped the team adhere to high standards throughout the life of the project		
Communication	**You**	**Your team**
11. Carefully prepared presentation materials in advance		
12. Practiced making your presentation before giving it		
13. Provided constructive feedback about what and how material was presented		
14. Effectively used available technology to enhance the quality of your presentation		
15. Prepared a thorough written final report that was clear and easy to follow		
Collaboration	**You**	**Your team**
16. Ensured that all team members participated in the final presentation		
17. Shared credit for team success and performance		
18. Carefully reflected on the team's overall effectiveness in terms of collaboration		
19. Helped others to improve their collaboration skills		
20. Celebrated team successes and accomplishments		

BIBLIOGRAPHY

BARKER, J. A. *Paradigms: The Business of Discovering the Future*. HarperCollins, New York, 1996.

DHILLON, B. S. *Engineering Design—A Modern Approach*. Irwin, Chicago, 1996.

GUNS, B. *The Faster Learning Organization: Gain and Sustain the Competitive Advantage*. Jossey-Bass, San Francisco, 1996.

ISBELL, D., HARDIN, M., and UNDERWOOD, J. *Mars Climate Orbiter Team Finds Likely Cause of Loss*. NASA Press Release 99-113 (http://mars.jpl.nasa.gov/msp98).

KERZNER, H. *Project Management: A Systems Approach to Planning, Scheduling and Controlling*. 4th ed. Van Nostrand Reinhold, New York, 1996.

LAWBAUGH, W., HOBAN, F., and HOFFMAN, E. *Readings in Program Control*. NASA SP-6103, Washington, DC, 1994.

SAGE, A. *Systems Engineering*. John Wiley & Sons, New York, 1992.

SHISKO, R. *NASA Systems Engineering Handbook*. NASA SP-6105, Washington, DC, 1995.

Syllabus. *New Directions in Education Technology*, 13:1, August 1999.

VOLAND, G. *Engineering by Design*. Addison-Wesley, Reading, MA, 1999.

BEHAVIORAL CHECKLIST SUMMARY FORMS

INSTRUCTIONS

Use the forms in this Appendix to help you summarize the ratings you made using the behavioral checklists at the end of each chapter. You will find three forms. Use the first one to calculate your average self-ratings for each skill area and design phase. Use the second form to calculate your average ratings of your team. The third page is a development planning form. Use it to summarize your strengths and to identify some specific ways you and your team members might improve your effectiveness in the future.

CALCULATING AVERAGE RATINGS

Use the form entitled *Average Self-Ratings* to obtain an overview of how you rated yourself:

1. Begin by referring to the behavioral checklists at the end of four chapters. The checklists can be found on pages 70, 131, 172 and 209. For each checklist, sum your self-ratings (Those ratings under the heading "you") for each skill area (decision making, project management, communication and collaboration and all four areas combined).
2. Record this number in the appropriate boxes on the line beneath the word "sum."
3. Calculate your average ratings by dividing each sum by the number specified in each box. Record your average in the appropriate boxes on the line beneath the word "average."
4. Use the last row of the matrix to calculate an average rating for each skill area across all four phases.

Use the form entitled *Average Team Ratings* to summarize how you rated your team. Complete the form by following the same procedures outlined above. Only in this case, use the ratings under the heading "your team."

COMPLETING THE DEVELOPMENT FORMS

Use these forms to record your areas of strength as well as some ways you might improve. There are two copies of the form, one for your self-ratings and one for your ratings of your team.

1. Begin by reviewing your average ratings on the *Average Self-Rating* form. In general, consider any skill area with an average rating of 3.5 or higher to be an area of strength. Obviously, the higher your ratings, the more likely it is you view the skill as a strength.
2. As you review your average ratings, try to assess their consistency across various design phases. Did you tend to rate yourself more or less the same in terms of a particular skill area? Alternatively, are your ratings more consistent across skills within the same phase of the design process? If there are such patterns, ask yourself why this might be the case. Maybe you are particularly effective in one skill area or another. Perhaps you were particularly comfortable with your abilities to contribute during a particular design phase or maybe you found yourself overwhelmed by other schoolwork during a certain phase of your design project. In any event, contemplating some of these issues can help you in terms of better understanding your strengths and improvement opportunities.
3. Use the space provided to describe your strengths and improvement areas. When doing so, be as specific as possible. It may help to refer back to individual items from the various checklists you completed. You will have a greater chance of improving if you can define your improvement objectives in specific terms. For instance, instead of stating: "*I want to improve my conflict management skills*," specify exactly what you will do differently (e.g., "*During my next team project I will focus on conveying my understanding of another person's point of view before offering a conflicting opinion*.") Refer to the appropriate sections of this text to help you write more specific development objectives. In addition, seek suggestions from your professor, design team members, and friends.

Once you have completed a development form based on your average self-ratings, do the same for your team ratings. Ideally, all team members should complete this form and share copies with one another.

Average Self Ratings

Design phase	Decision making	Project management	Communication	Collaboration	Average rating for each phase
Defining the problem	Sum Average _____/5 = ___	Sum Average _____/5 = ___	Sum Average _____/5 = ___	Sum Average _____/5 = ___	Sum Average _____/20 = ___
Formulating solutions	Sum Average _____/5 = ___	Sum Average _____/5 = ___	Sum Average _____/4 = ___	Sum Average _____/5 = ___	Sum Average _____/19=___
Developing models and prototypes	Sum Average _____/5 = ___	Sum Average _____/5 = ___	Sum Average _____/5 = ___	Sum Average _____/5 = ___	Sum Average _____/20 =___
Presenting and implementing the design	Sum Average _____/5 = ___	Sum Average _____/5 = ___	Sum Average _____/5 = ___	Sum Average _____/5 = ___	Sum Average _____/20 = ___
Average rating for each skill area	Sum Average _____/20 = ___	Sum Average _____/20 = ___	Sum Average _____/19 = ___	Sum Average _____/20 = ___	

Average Team Ratings

Design phase	Decision making	Project management	Communication	Collaboration	Average rating for each phase
Defining the problem	Sum Average _____/5 = ___	Sum Average _____/5 = ___	Sum Average _____/5 = ___	Sum Average _____/5 = ___	Sum Average _____/20 = ___
Formulating solutions	Sum Average _____/5 = ___	Sum Average _____/5 = ___	Sum Average _____/4 = ___	Sum Average _____/5 = ___	Sum Average _____/19=___
Developing models and prototypes	Sum Average _____/5 = ___	Sum Average _____/5 = ___	Sum Average _____/5 = ___	Sum Average _____/5 = ___	Sum Average _____/20 =___
Presenting and implementing the design	Sum Average _____/5 = ___	Sum Average _____/5 = ___	Sum Average _____/5 = ___	Sum Average _____/5 = ___	Sum Average _____/20 = ___
Average rating for each skill area	Sum Average _____/20 = ___	Sum Average _____/20 = ___	Sum Average _____/19 = ___	Sum Average _____/20 = ___	

Development Planning Form (Self)

Your name: ______________________________

1. Some of the key things I learned about the engineering design process during this project are:

2. Some issues and topics about which I would like to learn more are:

3. Some examples of ways I was particularly effective during this project are:

4. Some specific things I could do to be even more effective in the future are:

5. To help me improve my skills I would appreciate it if others would:

Development Planning Form (Team)

Your team's name: ______________________________

1. Some specific skill areas/behaviors that we as a team were good at are:

2. Some specific skill areas/behaviors in which we could improve as a team are:

3. The biggest challenges we faced and overcame as a team were:

4. Some specific things I recommend this team do in the future to be more effective are:

INDEX

Page references followed by f or t refer to figures or tables, respectively.